Rife's World of Electromedicine

Rife's World of Electromedicine

The Story, the Corruption and the Promise

By **Barry Lynes**

Author of **The Cancer Cure That Worked**

BioMed Publishing Group

BioMed Publishing Group
P.O. Box 9012
South Lake Tahoe, CA 96158
www.BioMedPublishers.com

Copyright 2009 by Barry Lynes
ISBN 10: 0-9763797-9-1
ISBN 13: 978-0-9763797-9-9

For related books and DVDs, visit www.biomedpublishers.com.

Disclaimer

The opinions expressed in this book are solely those of the author and do not represent the viewpoint of BioMed Publishing Group.

This book is not intended as medical advice. It is also not intended to prevent, diagnose, treat or cure disease. Instead, the book is intended only to share the author's research, as would an investigative journalist. The book is provided for informational and educational purposes only, not as treatment instructions for any disease. Much of the book is a statement of opinion in areas where relevant facts are controversial or do not exist.

If you have a medical problem, please consult a licensed physician; this book is not a substitute for professional medical care. The statements in this book have not been evaluated by the FDA. The author is not a medical doctor or healthcare practitioner.

TABLE OF CONTENTS

Preface

In April of 1987 a book that I authored was published and quickly became an alternative health/science phenomenon because it exposed a monstrous scandal that "official medicine" and "official science" remained unwilling to recognize as documented truth.

The book, titled *The Cancer Cure That Worked—The Rife Report,* revealed the long-suppressed and censored cure for cancer as demonstrated by a 100% successful cancer clinical trial managed by a university medical school and five of America's most prominent physicians.

Now, more than two decades after the original book, the new book in the reader's hands explains what 20+ years of additional research and document retrieval by a dedicated grass roots group have discovered and unearthed. Some of the previously missing, key factors in the suppressed and censored cure for cancer were lost or even locked away in physicians' closets for decades in order to prevent censors from destroying overwhelming evidence.

This new book—*Rife's World of Electromedicine*—also provides focused information on a continuing crime and suppression of an entirely new science that medical and scientific officials are seeking to keep hidden.

Welcome to the dawn of a new world—*Rife's World.*

(Note: To learn more about Barry's book, The Cancer Cure That Worked, please see Appendix E)

Foreword:
The Story

The cause of cancer was discovered in the early 1930's. It was a virus-sized, mini-bacteria or "particle" that induced cells to become malignant and grow into tumors. The cancer microbe or particle was given the name BX by the brilliant scientist who discovered it: Royal Raymond Rife.

Laboratory verification of the cause of cancer was done hundreds of times with mice in order to be absolutely certain. The scientist who made the historic discovery then worked with five of America's most prominent physicians in order to test the cure for cancer in a well organized cancer clinical trial managed by a major university's medical school. The year of the cancer clinical trial was 1934.

Sixteen cancer patients competently diagnosed as hopelessly terminal were brought by ambulances twice a week to the cancer clinical trial location in La Jolla, California just north of San Diego. There they were treated with a revolutionary electro-medicine that painlessly, non-invasively destroyed only the cancer-causing microbe or particle named BX. After just three months of this breakthrough, science-based therapy, all of the cancer patients were diagnosed as clinically cured. Sufficient eye-witness statements in writing and documents by prominent scientists who measured the tumor shrinkage exist.

A profound, history-making event had occurred.

During the next 4-5 years, the original electronic technology was improved, tested, and used in continuing clinics in three locations—Los Angeles, Pasadena, and San Diego. A steadily growing case history record of cancer cures resulted.

Unfortunately, one of the successful cases was an 80-year-old man from Chicago, Illinois with a large facial cancer. But after treatment, he returned to Chicago with skin like "a new born baby's" on his face. His friends were astonished and word of the miracle treatment quickly spread. The head of the American Medical Association (AMA)—with headquarters in Chicago—soon found out what had happened in far-away San Diego.

Agents of the AMA suddenly showed up in San Diego. An engineer employed by the scientist who discovered the cause and the cure for cancer was quickly bribed. A lawyer from the AMA rapidly followed and—along with the bribe-taker—the two of them sabotaged the company that built the stunningly reliable and effective technology. Then in 1939, the AMA lawyer and the bribe-taker initiated a lawsuit in order to steal one of medical history's greatest discoveries. The trial lasted three months (June-August, 1939).

The AMA lost its criminal attempt to grab ownership of the new science and the great, transformative medical treatment. Therefore, in order to hide its wicked act, the AMA imposed a massive suppression and censorship on both the new science and the new medicine.

Millions of people with cancer suffered and died in subsequent decades because of the American Medical Association's monstrous deed and crime against humanity.

Evidence exists to prove the accusations set forth here. The evidence includes eye witness reports, court transcripts, an article describing the new science in the Smithsonian Institute, published micro-photographs, and other historical details.

Many 21st century, modem day, educated professionals are aware of the facts just summarized. This includes editors at major magazines and newspapers, an investigative show's producer at a major television network, book publishers, literary agents, etc. Unfortunately, the world class story apparently terrifies them. They run from it. They dare not be the one who breaks the story in the mainstream media despite the sufficient documentation and factual evidence. They apparently fear for their own personal careers, and thus choose to play it safe.

There is a deep moral principle that is being thwarted by such career-based, self-censoring behavior combined with fear.

Facts are facts. Someday the whole, awful truth will emerge into a glaring public spotlight. Once a major breakthrough in science occurs, its replication and further development is assured.

The American public is not likely to forget or condone such outrageous, self-serving behavior by its supposed professional watchdogs. Hiding the truth and participating in a cover-up of this magnitude resembles the *New York Times* editor sadly admitting that his newspaper missed clear evidence of the Nazi Holocaust! And now, something equally terrible stains the pages of history.

Chapter 1
The Promise: Electromedicine, the Science of Tomorrow

The science is breathtakingly simple. Disease conditions can be identified by unique wavelengths that define colors. Scientists estimate that humans can recognize 10 million shades of colors! (*National Geographic Magazine*, October 2001, page 17.)

Then, once the microbe that caused the patient's disease condition is identified by its color, it can be painlessly and non-invasively destroyed by a specific radio frequency that destroys only that disease-causing micro-organism. That two-step process should not be hard to grasp.

Most invasive procedures and surgeries could be relegated to the dustbin of history.

As the renowned physician of his time Dr. George Crile adventurously prophesied during a medical conference speech in 1933 (reported on page 1 of the *New York Times* newspaper for October 9, 1933):

> *The medical man of the future would tune in on the living body as one does now the ordinary radio. By listening in ...*

> *the physician-radio-engineer of the future would thus be enabled to tell ... whether they were playing a melody of health or ... signaling an SOS.*

In the year 1999, Rutgers University Press boldly published *Evaluating Alternative Cancer Therapies* by Donald J. Hess. The book included this priceless declaration:

> *The future of medicine is in light and electromagnetic fields ... Probably the guy who knew the most was Royal Raymond Rife.*

The *Miami Herald* newspaper reported:

> *Rife's ... discovery went virtually unnoticed for more than fifty years until ... Lynes' book about Rife was published.*

Dr. Peter Macomber, M.D., Harvard trained pathologist and former Assistant Chief of Experimental Pathology at Walter Reed Army Institute of Pathology, Washington, D.C., explained in a statement published by the *Townsend Letter for Doctors and Patients* magazine:

> *Royal Rife ... designed ... electrical machines which he could tune to certain frequencies and ... cause the (cancer microbe) to disintegrate.*

> *Apparently the bulk of modern cancer research and therapy has been misdirected at studying and trying to remove or destroy the cancer cell rather than the cancer (microbe) ... We must now rapidly redirect research and therapeutic efforts toward the cancer (microbe).*

The *San Diego Evening Tribune* newspaper of May 6, 1938 reported:

> *Rife thinks that the lethal frequencies for various disease organisms are ... coordinates of frequencies existing in the organisms themselves ... Brief exposure to the frequencies ... brings the fatal reactions. In some organisms, it happens in*

seconds ... Now ... the ray can be turned upon that organism with the assurance that the organism will be killed.

In the 21st century, *Everyday Practical Electronics* magazine, a mainstream media publication in England, published an incredible overview of Royal Rife's scientific discoveries titled "The End to All Disease." The article included the significant statement:

Rife ... hit upon the idea of illuminating the object with two polarized beams of monochromatic light ... He didn't need a stain, the light itself became the stain ... The implications were profound. He had seemingly found a way to selectively destroy ... pathogens without damaging human cells.

In the year 1935, Royal Rife provided the following, clear explanation and description of his history-making discovery and how the new technology worked:

The basic principle of this device is the control of a desired frequency. These frequencies varying upon the organism being treated ... The frequency with its carrier wave is transmitted into an output tube similar to a standard x-ray tube, but filled with a different inert gas. This tube acts as a directional antenna. The importance in the variable control of these frequencies is that each pathogenic organism being treated is of a different chemical constituency, the consequence being they carry a different molecular vibratory rate. Each one in turn under these conditions requires a different frequency or vibratory rate to destroy.

One of Royal Rife's colleagues stated in a mid-1950's taped interview:

He finally got these cultures on the slide. He could look through this thing and you could see them swimming around absolutely motile and active. Then he'd say, 'Watch that.' He'd go turn on the frequency lamps. When it got to a certain frequency, he'd release the whole doggone flood of power into the room. The doggone little things would die instantly.

I like Roy Rife ... this tremendous thing that he built which we call the Rife ray machine ... Roy was one of the most gentle, genteel, self-effacing, moral men I ever met. Not once in all the years I was going over there to the lab, and that was approximately 30 years, did I ever hear him say one word out of place.

All the doctors used to beat a path to Rife's lab door and that was a beautiful lab ... The equipment was just exactly right ... the atmosphere could not be duplicated.

One modern internet site bluntly proclaims and predicts:

It is a fair statement that Rife practically developed bioelectric medicine himself ... One day the name of Royal Raymond Rife may ascend to its rightful place as the giant of modern medical science.

—www.healthresearchbooks.com/articles/rife.htm

Even the prestigious Smithsonian Institution at one time championed Royal Rife's phenomenal medical discovery and new science. The *Smithsonian Annual Report* told its readers what now is cruelly kept from public knowledge and public attention. The Smithsonian Institute dared to publish:

... disease organisms such as those of tuberculosis, cancer, sarcoma ... typhoid ... and others may be observed to succumb when exposed to certain lethal frequencies peculiar to each individual organism, and directed upon them by rays conveying a wide range of waves.

High U.S. government scientific officials and scientific institutions committed an historic dereliction. They made mindless, self-serving, unscientific errors and policy decisions that kept the clinical-trial-demonstrated cure for cancer a taboo area for researchers until modern times. Fresh, related discoveries of historical documents in the 21st century have resulted in a rapidly evolving grass roots movement and partially successful renaissance of this tragically lost science and medical technology. The resurgence of interest and global aware-ness of Royal Rife's accomplishment grows with each passing year toward an inevitable day of international recognition.

Soon it will be time for the press, publishers, and people of conscience to demand straight answers about some very ugly, long hidden truths. As Peter Andrews declared in an *American Heritage* magazine article:

> *A newspaper is a call to action. It cannot know all the answers, but it can ask the right questions.*

Chapter 2
Rife's Legacy to Humanity

The enormous importance of Royal Rife's discoveries was obvious. That is now, from an historic perspective, an indisputable fact. No debate needed.

However, at the time Royal Rife created an entirely new science using (1) color as the means of identifying the unique microbe responsible for a specific disease, and (2) the frequency device that would kill only the specific targeted pathogen (germ or microbe), neither Royal Rife nor his powerful supporters had the slightest awareness of the obstacles or the greedy opposition determined to prevent Rife's new science from being developed and publicly accepted.

How fragile a new science at its birth truly is. So many things can go wrong—any of which can stop the unfolding development process at a crucial, early stage.

Royal Rife was tragically blindsided by a payoff-taking person within his own circle of associates and by a powerful, tyrannical human monster in the outside world. This malefactor of medicine was determined to steal the history-making discoveries for himself and his medical organization.

Still, what an accomplishment Royal Rife's discoveries were! Now, in the 21st century, a growing awareness of what Royal Rife did brings closer and closer an inevitable recognition of the new science.

In the 21st century, the editor of the British mainstream magazine *Everyday Practical Electronics* boldly communicated to his readers:

> *The level of interest in this material ... has been amazing ... We hope that by giving exposure to the original work of Rife it will encourage a more open-minded approach by those in the medical profession ... for too long powerful organizations with vested interests have suppressed development and research in this area.*

But back when Royal Rife was carefully bringing forth the new science with its incredibly simple 2-part base (color identification of the disease-causing microbe followed by electronic frequency destruction of the "germ"), the challenge was not the cruel intrigues of medical politicians, but rather unlocking the secrets of nature through the slow, careful scientific process.

Royal Rife's own words provide instructive insights for future researchers and those who dare to finance this incredible new science:

> *... owing to the minuteness of these particles, it is impossible to stain them with any known method or technique using acid or aniline stains; hence a substitute stain was found ... a frequency of light that coordinates with the chemical constituents of the particle or micro-organism under observation.* Royal Rife, History of the Development of a Successful Treatment for Cancer, 1953.

> *We can make a positive diagnosis of any disease from which the filtered ... forms can be observed ... checked 181 times against a color dictionary to see if the color of the various forms would change a shade or two but they have been found to be very constant.* Royal Rife, "Cultivation," 1958.

> *An electronic frequency to devitalize the organism, the successful treatment of cancer. No tissue is destroyed. No*

pain is felt. No sensation is noticed. A tube lights up and 3 minutes later the treatment is completed ... The body then recovers itself naturally from the toxic effect ... Several disease forms may be treated simultaneously. Royal Rife, History of the Development of a Successful Treatment for Cancer, 1953.

We sincerely believe ... that BX is the primary cause of cancer ... We have used Koch's postulates in our methods of recovery ... this has been repeated hundreds of times proving to our own satisfaction that BX ... is the cause of malignancy. Royal Rife, History of the Development of a Successful Treatment for Cancer, 1953.

'Koch Postulates,' as they became known, stated that the organism had to be present in every case of the disease, that it could be cultured for several generations (thus establishing that the cause of the disease was a living entity), and that, when introduced into a healthy subject, the organism should induce the same disease ... With a few exceptions, Koch's postulates still form the basis of bacterial diagnosis. Scientists & Inventors, Anthony Feldman and Peter Ford, 1979, page 199.

The first clinical work on cancer ... was made under a Special Medical Research Committee of the University of Southern California ... 16 cases were treated for many types of cancer. After 3 months ... these so-called hopeless cases were signed off as clinically cured by the staff of five medical doctors. Royal Rife, History of the Development of a Successful Treatment for Cancer, 1953.

According to Royal Rife's laboratory notes from the 1930s, the cancer causing microbe was so small that if 500,000 of them were put side by side, the total size would be only one inch.

Chapter 3
The Corruption: Treachery by Government Health Officials

Royal Rife's easily understood and easily verified scientific breakthroughs were not welcomed by high U.S. government health officials and policy makers. Why? Because Royal Rife's history-making, magnificent discoveries threatened a very comfortable status quo world for the bureaucrats, civil servants, and regulators who lived there.

Simply put, ugly as the reality certainly is, the government-employed health lawmakers didn't seem to care that their contemptible policies, habits, and procedures were killing people.

They had nice, secure lifetime employment with fully paid health benefits, weekends off, and good pensions. They lived comfortable lives. And from their protected world they betrayed the public trust. Royal Rife's incredible, breakthrough science and magnificent medical method of healing vicious diseases didn't interest them. Some of them even openly acknowledged to journalists and were quoted in newspaper articles that they were pawns of the drug industry. They openly admitted it! Royal Rife's discoveries were a dangerous threat to their personal well being and their personal career advancement. Therefore, Royal Rife's modern day, 21st century revival had to be ignored, opposed or attacked.

This behavior was a simple, if depraved truth. Even if Royal Rife's clinical trial-demonstrated cure for cancer had overwhelming documentation (which it does), the bureaucrats' fanaticism against it persisted. The cost was huge suffering and death for countless American cancer patients.

Were the regulators and health officials just dumb? No. There was something much more sinister in their behavior and misdeeds that caused their fervent dislike for Royal Rife's history-making achievement. How could a responsible society have permitted the two primary health agencies of FDA (Food and Drug Administration) and NCI (National Cancer Institute) to remain on such a wrong course for so many decades? Honest testing of Royal Rife's new science was never done by FDA and NCI officials. Never! Access to Royal Rife's revolutionary and successful medical therapy was never offered or allowed to cancer patients, even if the patients were dying.

This was unconscionable policy.

Certainly there existed enough solid evidence in this unjust, injurious, immoral, and ongoing deceit to justify some broad Congressional investigations. And also some aggressive pursuits of long hidden, factual truths by a newly aroused press corps.

The problem was so outrageous and fundamental that it could be addressed in the context of a huge scientific leap that had been prevented by a group of government scientists with mediocrity being their outstanding attribute. The issue can be summarized in a straightforward, blunt, and stunningly horrifying verbal picture:

Royal Rife developed a microbe and pathology classification system of "germs" that someday probably will be recognized as rivaling the

Periodic Table of Elements, the fundamental basis of modem chemistry and physics.

A Periodic Table for disease-causing micro-organisms was the scientific foundation for the 100% successful cancer clinical trial. Therefore, with the wonderful treatment already a medical taboo, an entirely new scientific field simultaneously became taboo. The dimensions of this tragedy expand geometrically as different details become known and are examined.

Imagine destroying, hiding, or ignoring a scientific discovery of that magnitude.

In fact, Royal Rife's color-designating diagram for all pathogens—disease-causing germs—had actually been compared by knowledgeable people to be the equivalent of the Periodic Table of Elements.

In a taped interview that occurred in the middle 1970's decade, an M.D. university professor and pathologist heard the following from a person familiar with Royal Rife's masterful diagram:

> *They were like the Periodical Charts for Elements. His (Royal Rife's) charts were for micro-organisms ... all the pathogens fit into one or another of these columns. All the pathogens ... of various types ... up through an evolutionary type ladder ... all human pathogens fell onto this chart.*

Chapter 4
Microscope Heaven

The history of microscopes is not complicated.

In 1674, a Dutch merchant named Antoni van Leeuwenhoek looked through some beautiful lenses he had created and discovered bacteria. The world of microscopes and microbiology had been born!

Leeuwenhoek spent the following decades sending sketches of his bacteria discoveries to the Royal Society in England. The scientific world of micro-organisms began to be slowly, carefully documented.

The first evidence that we are surrounded by multitudes of microbes was provided by observations made by Antoni van Leeuwenhoek with primitive microscopes in 1674 ... As a draper ... He had the ability to make small lenses—only about 1 millimeter in diameter—of superb quality ... these simple microscopes revealed to Leeuwenhoek clear images of very small objects, magnified as many as 300 times or more. Howard Gest, The World of Microbes. 1987, pages 2, 3.

... he gazed into a world which until then had been wholly unknown ... What Leeuwenhoek unveiled in the world of the miniature was scarcely less important than the discovery of America ... this Columbus of the microcosm. Gerhard

Venzmer, Five Thousand Years of Medicine. 1968,1972, pages 178, 179.

For the next two centuries, microscope creation advanced as various inventors contributed to its evolution. And then, in the 19th century, another genius appeared. He was a German physicist named Ernst Abbe who lived from 1840 to 1905. In the 1880's, Abbe virtually created by himself the modern light microscope still used today in the 21st century. A light microscope is also called an optical microscope. Ernst Abbe was co-founder of the world renowned Zeiss Optics company.

> ... the manufacture of microscopes remained an art rather than a science until Carl Zeiss of Jena decided to form a partnership with a young physicist called Ernst Abbe (1840-1905). From 1883 to 1895 Abbe solved the basic problems of light microscopy ... There, in 1883, were made the first lenses corrected for both chromatic and spherical aberration, and by the turn of the century microbiologists were provided with light microscopes similar to those used today. Morris Solotorovsky, Three Centuries of Microbiology. 1965, 1970, page 4.

Ernst Abbe mentored a successor in genius-level microscopy named Hans Luckel. Hans Luckel in turn mentored a microscope prodigy named Royal Rife (1888-1971). Royal Rife, born in mid-America Omaha, Nebraska eventually settled in San Diego, California where he made microbiology discoveries and invented a variety of scientific instruments at a "Da Vinci level awareness"—several decades ahead of his time.

In 1931, Royal Rife's name was first introduced to a national readership.

> We were in the laboratory of R. R. Rife at San Diego, California ... For ten years he has worked to capture in motion pictures what the eye sees through the most powerful

> *microscopes ... sixteen of the finest quartz lenses obtainable*
> *... magnify the dimensions of each germ 12,000 times.*
> *Designed by Rife himself, this apparatus is one of the most*
> *powerful microscopes in the world; its magnification*
> *compares with the 2,000 diameter enlargement of*
> *microscopes commonly used in research laboratories.*
> Popular Science magazine, June 1931.

Royal Rife began and continued to create and improve his revolutionary microscope design and by the middle 1930's he had created his masterpiece microscope. He called it the Universal Microscope. Its magnification and resolution capabilities were astounding.

Twenty years later, in the mid-1950's, Royal Rife provided the following explanation of how his microscope worked and how the linkage between a specific disease-causing germ's color was obtained. The description is a bit technical, but if the general reader simply stays with the single color (monochromatic) notion, he or she should grasp the beauty and absolute magnificence of Royal Rife's genius-level invention.

> *When the quartz prisms on the universal microscope ... are*
> *rotated in opposite directions, they serve to bend the*
> *transmitted beams of light at variable angles of incidence*
> *while, at the same time, a spectrum is projected up into the*
> *axis of the microscope, or rather a small portion of a*
> *spectrum since a part of a band of color is visible at any one*
> *time. However, it is possible to proceed in this way from one*
> *end of the spectrum to the other, going all the way from the*
> *infrared to the ultraviolet. Now, when that portion of the*
> *spectrum is reached in which both the organism and the*
> *color band vibrate in exact accord, one with the other, a*
> *definite characteristic spectrum is emitted by the organism.*
> Royal Rife, "The Evolution of the Microscope," The
> AeROHRcrafter corporate newsletter, January 1954, Pages 11,
> 12.

In 1940 the *Los Angeles Times* published an article about Royal Rife's incredible microscope with an accompanying photograph of Rife and the "Universal."

> *Capable of magnifying an object 31,000 times, compared with the 1600 to 1700 times of the standard microscope, Rife's instrument had disclosed ... minute, deadly enemies of the human body which never before ... have been seen ... Unlike the standard microscope, the image does not pass through free air in a hollow tube, with resultant distortion. It is conveyed, instead, zigzag fashion through quartz blocks and prisms along the optical path.* Los Angeles Times, June 26, 1940, Part II, page 13.

The Electron Microscope came along in the late 1930's also. It eventually produced magnifications much greater than Royal Rife's Universal Microscope. However, Rife was not impressed because the result was dead micro-organisms—the electron microscope was incapable of observing living micro-organisms. How can bacterial life truly be understood without observing it in action, alive? This has been the major flaw of the last 75 years of microbiology, followed by the inevitable errors in diagnosing and treating disease. Rife's microscope instead revealed living, changing micro-organisms along with a color classification of disease-causing microbes.

Two quotations complete this brief overview of microscope history. These quotations should provide the attentive reader with a general understanding of why Royal Rife's microscope approach could have taken all subsequent microbiology research and 20th century medicine in a much more productive direction.

> *... the filterable ... forms have been demonstrated to be the causative agents of disease and the theoretical and applied biology cannot be denied ... The electron microscope will not show these filterable forms as live entities because as all scientists know, that when micro-organisms are prepared*

for observation, gelatin slides, and various stains are used to show obscure forms. Royal Rife, "Cultivation," 1958.

These optical microscopes had one major advantage over the modern electron microscopes. They allowed the study of live specimens. An electron microscope operates in a high-vacuum chamber and bombards the specimen with high energy electrons; not much can survive that. "The End to All Disease," Everyday Practical Electronics (British mainstream magazine), April 2001.

Chapter 5
BX: The Cancer Causing Particle

What Royal Rife saw in his marvelous Universal Microscope and isolated as the cause of cancer was a microbe. The cancer microbe. The reader needs to keep that single, simple idea in mind. It is crucial to understanding the tremendous implications of Rife's great accomplishment for the medical science of tomorrow.

Now what is a microbe?

> *"Microbe" is merely a convenient name for any of hundreds of thousands of species of microscopic organisms that flourish on earth; the most numerous are the ones we call bacteria ... they are structurally the simplest, lacking the cell nucleus found in other microbes. Most reproduce by fission. They multiply by dividing.* Thomas Y. Canby, "Bacteria," National Geographic magazine, August 1993, pages 41, 42.

Now what does "filter-passing" form mean?

A filter catches bacteria because of bacteria's size. What gets through the filter is so small that a standard light microscope or optical microscope won't reveal it to the scientist looking through a standard microscope.

But in the early 1930's Royal Rife, using his super microscope, could view and study the living, filter-passing forms. And it was there that he found the "particle" that causes cancer. He called it the "BX."

Using mice, Royal Rife thoroughly confirmed his discovery over 300 times in his laboratory. The scientific laws of proof known as Koch's Postulates were used in order to be absolutely certain before Royal Rife attempted any clinical experiments in curing cancer with humans.

Royal Rife's laboratory notes for November 20, 1932 contain the first written description of the cancer microbe's characteristics. Rife later wrote, "BX ... will produce in the experimental animal a typical tumor with all the pathology of true neoplastic tissue, from which we can again recover the 'BX' micro-organism. This complete process has been duplicated over 300 times with identical and positive results."

Royal Rife described the cancer microbe as follows in 1933:

> *The 'BX' is ... filterable ... It is a small ovoid (egg-shaped) granule, highly plastic, and visible only with monochromatic light. The color is purple red. The x ray has no influence on the organism. The application of the oscillative ray should completely destroy the BX in the malignant tissue. Thus a new field is open to scientific investigators.* Royal Rife, paper dated 1933.

Royal Rife described BX as a "particle." In his 1953 report to the nation's cancer authorities (that they denounced), Royal Rife stated:

> *... owing to the minuteness of these particles ... a frequency of light ... that coordinates with the chemical constituents of the particle (was used to stain them).* Royal Rife, History of the Development of a Successful Treatment for Cancer, 1953.

Royal Rife's scientific reach was matched by the clarity of his writing and a humility that allowed him to include later generations of scientists who would follow him into the great unknown where this incredible new science would be further developed.

> *... to look upon disease-causing organisms in their true form ... We sincerely hope that our efforts ... will stimulate and create a desire in the minds of other workers to carry on in the broad and inviting field before us, one which presents a work so vital and essential for the benefit of all mankind.* Royal Rife, "The Evolution of the Microscope," The AeROHRcrafter corporation Newsletter. January 1954, pages 11, 12.

In 1944 the world renowned Smithsonian Institute of Washington, D.C. published its annual report. The report contained an article describing the scientific wizard Royal Rife's astounding discoveries and inventions. Unfortunately, World War II was blazing in its global horror and the significance of the Smithsonian Institution's historic announcement was overlooked. But the Smithsonian Report now stands forever as a document for history. It contained the following timeless description.

> *More than 20,000 laboratory cultures of carcinoma were grown and studied over a period of 7 years ... Then, in 1932 ... a new approach to the problem ... purplish-red, filterable forms measuring less than one-twentieth of a micron in dimension, were observed ... the BX.* Annual Report, The Smithsonian Institution, 1944, U.S. Gov. Printing Office, 1945, Publication 3776.

By destroying the BX "particle" in the 16 cancer patients of the 1934 cancer clinical trial, Royal Rife demonstrated the truth of his scientific discovery as well as the effectiveness of the incredible revolutionary medical treatment using radio frequencies.

In 1940, Arthur Yale, M.D. read the following to the California State Homeopathic Medical Society. The written report was subsequently published by the Medical Society.

> *For seventeen years Mr. Rife has succeeded in finding a vibratory rate which will kill the different invading organisms of the body. Having used this apparatus for almost two years, the writer has had the satisfaction of witnessing the disappearance of every malignant growth where the patient has remained under treatment ...* Arthur W. Yale, M.D., "Cancer," *Pacific Coast Journal of Homeopathy,* July 1940.

Fifty-seven years after Dr. Yale's report, another serious voice described what the modem day, 21st century cancer "experts" continued to evade.

> *The key to fighting cancer and other diseases successfully lies in the potential of destroying the pathogenic microorganisms and getting the body to excrete them. The microbial activity is the basis for the growth process of the tumor.*
>
> *The remedies can heal the individual at any stage in the growth process if the microbial progression is reversed.* Michael Coyle, Advanced Microscopy for Nutritional Evaluation and Correction, 1997.

Traditional scientists had known for decades that living microbes smaller than bacteria had passed through bacteria-stopping filters. But these living "particles" or "dwarf bacteria" took them into Royal Rife's territory. And traditional scientists sure did not want to venture there—such an indiscretion could be very dangerous for their personal careers. Therefore, they mainly just avoided the subject.

In other words, it was safer to investigate areas where the Club was comfortable. (The Club was comprised of National Cancer Institute "experts" who controlled the massive projects funded by multi-million dollar grants.)

However, the political reality of serious money and cancer center credentials could not hide the disturbing fact that serious scientific literature had acknowledged (for decades) that something existed in the off limits territory where Royal Rife had discovered BX, the cause of cancer.

Compare the following description from a mid-1960's book with what Royal Rife was doing 30 years earlier, and curing cancer as a result.

> *... there appear granulations, called 'dwarf forms,' of vary small dimensions, measuring in fact a millionth of a millimeter ... They seem to be rather fragile and possess a very special property, due to their smallness, of being able to go through the walls of filters which retain normal bacteria.* Alan Deaunay, The World of Microbes. 1965, page 43.

Still, the Club's priorities remained directed at treating cancer cells and cancer tumors. Cancer-causing microbes or living "particles" of virus size threatened all sorts of entrenched interest groups.

Royal Rife's long ignored discoveries were controversial, off limits, and definitely not welcomed by those managing orthodox cancer centers. The high paid "experts" pouring various chemotherapy concoctions into the veins of cancer patients kept "alternative" medicine restricted to safe "fringe" or "complementary" areas. The cancer experts at the National Cancer Institute (NCI) and the Food and Drug Administration (FDA) have refused to allow anyone to investigate Royal Rife's cancer particle-based, scientific, 100% successful clinical trial approach to curing cancer. Yet the chemotherapy that the NCI and FDA have defended and promoted had one small, obvious problem. It killed cells, not the underlying cause of cancer. And therefore the NCI and FDA cautious "cancer experts" and obstructive regulators forced an enormous and unnecessary death toll upon American citizens who had cancer.

> *Chemotherapy ... is a blunt instrument. The drugs attack every rapidly dividing cell in the body.* The Economist magazine, October 28th, 2006, page 90.

Chapter 6
The "Gold Standard" Scam

The wrath of the cancer "experts" can be quite intimidating. It usually scares people into obedience and passivity. And that is the way the cancer medical moguls want the existing system to be and to remain.

The officials who essentially manage the American nation's cancer program are primarily relying on the cancer clinical trial procedure and accompanying statistical results as the "Gold Standard" for what they endorse and permit to be used in treating cancer.

Unfortunately, such an emphasis obstructs simple, common sense solutions to curing cancer. It can produce such miscreant, absurd nonsense as a drug that extends life for only 6 months (the duration of a clinical trial) to receive unrestrained endorsement. People who have intelligence and common sense often cannot comprehend such a notion.

For those who have devoted their careers to procedures, regulations, analysis of numbers, and so forth, all this organized busy work makes perfectly logical sense. They obviously benefit enormously from it. Year-after-year and decade-after-decade new, experimental drugs or procedures arrive. Cancer patient volunteers are organized and the show begins anew.

Fortunately, there have been isolated voices that have recognized a scam exists. These few, but wise folks have exposed the rotten system for what it is—scientism. Scientism is NOT science, but rather a perversion of science. It is a sort of theology or belief system that causes enormous damage to cancer patients while benefiting the cancer system's hirelings or mercenaries.

Still, the National Cancer Institute (NCI) always seems to have a clever expert whom it trots out whenever a credentialed puppet is needed to deceive the public, the press, and the easily duped Congressional oversight committee members.

Dr. Peter Greenwald of NCI made the Gold Standard nonsense quite explicit in a 1993 statement:

> *The gold standard of evidence in cancer research is clinical trials.* Reported by the Associated Press and published in the Los Angeles Times newspaper, March 15,1993.

Of course the 100% successful cancer clinical trial of 1934 using Royal Rife's breakthrough science and revolutionary electro-medicine treatment to destroy the microbe that causes cancer doesn't qualify for the gold standard of evidence needed. It was, after all, non-drug!

The gold standard is for drugs—complicated, pharmaceutical procedures and drawn out months of analyzing statistical results.

The misjudgement represented by this scam is so outrageous that its general acceptance is baffling. Unless one recognizes an underlying motive.

> *(It is) not our (FDA) policy to jeopardize the financial interests of the pharmaceutical companies.* Testimony of the FDA Commissioner to Congress. From G. E. Griffin, World Without Cancer, page 451.

Ah, the riddle solved! And that immediately returns us to the clinical trial gold standard scam.

> *If the controlled clinical trial were recognized for what it is— a mainstay of drug-industry and medical-industrial-complex monopoly—the antitrust impulse in American society would be directed against it today.*
>
> *The entire $550 billion American medical industry revolves, to some degree, around the controlled clinical trial. But because the theory is defective, the enormous superstructure erected upon it is equally shaky.* Harris L. Coulter, The Controlled Clinical Trial, 1991.

The fallacy of a clinical trial "gold standard" finally came to the attention of the medical scholar Robert J. Houston. He totally obliterated it with the following comment. Unfortunately, the lazy and timid media would not report such disturbing news to the American people.

> *In cancer, case studies have a greater degree of validity than in other diseases. In cancer the rate of spontaneous remission is extremely low, so low that it is virtually zero. Therefore, if you have just a few cases, basically if you have two cases, you have something that is solid. So I consider what is being dismissed as anecdotal evidence to be, in cancer, actually an impressive area of evidence, because you have much more detail in the case studies than you can in a clinical trial.* Robert J. Houston, Repression and Reform in the Evaluation of Alternative Cancer Therapies.

A deep sickness, incompetence, and corruption pervade orthodox approaches, conventional cancer treatment, the mindsets, and the arrogance of oncologists.

Two decades ago, a dedicated and soft-spoken M.D. named Raymond Keith Brown exposed "medical scientism" for the fraud that it is.

Medical scientism ignores the empirical approaches of traditional medicine. It discounts clinical observation and enshrines statistical analysis, double blind studies and the prevailing consensus of opinion.

Then there is the overt bribe element that has become an essential characteristic of modern day "gold standard" clinical trials.

The culture of clinical trials is plagued by conflicts of interest. Drug companies pay doctors handsomely—sometimes as much as $6,000 per patient—to test new drugs. U. S. News & World Report magazine, October 11, 1999, page 36.

The Food and Drug Administration (FDA) also joins NCI in the mess. Four pertinent quotations close this ugly issue with disturbing revelations.

Next year the drug industry will pay the FDA's Center for Drug Evaluation and Research $400 million in user fees. 'When you're getting paid directly from an industry you're supposed to be regulating, it takes quite a bit of edge out of the regulating,' says Sidney Wolfe, M. D. Patricia Barry, "Is the FDA Doing Its Job?" AARP Bulletin magazine, July-August 2007, page 16.

The Food and Drug Administration does very little to ensure the safety of the millions of people who participate in clinical trials, a federal investigation has found. Gardiner Harris, "Report Assails FDA," New York Times news service, September 28, 2007.

A chilling new report commissioned by the FDA's own Science Board ... "We were shocked at the appalling state of science at the FDA" says ... an adviser on the report. Reader's Digest April 2008, page 120.

... science, and scientists, can be bought—the public welfare be damned. Sharon Begley, Newsweek magazine, May 12, 2008, page 39.

Chapter 7
Beauty and the Beast

The beast was the American Medical Association (AMA). It killed Royal Rife's brilliant, demonstrated cure for cancer and Rife's discovery of an entirely new scientific field.

Beauty was the lost, elegant science and the many-colored microbes or "particles" that Royal Rife was the first human to see. Rife then classified the particles according to their color and what specific diseases each particle caused. It was a great advance for medicine and science.

In 1939, the AMA went to trial against Royal Rife. Their objective was simple—to steal Royal Rife's historic discoveries. The AMA had bribed an engineer who worked for Rife and provided the bribe taker with a top attorney in order to grab ownership of the new science.

Fortunately, the AMA lost its legal attempt to steal all of Rife's wonderful discoveries. Censorship and suppression of Royal Rife's work followed. However, the transcript of the court action provides an extraordinary record of what the AMA tried to pull off. The court transcripts expose forever the monstrous AMA action that kept the demonstrated cure for cancer and other diseases from being developed. A brief portion of the court transcript is now given.

The AMA Attorney: *"The machines coming from Dr. Rife would have Dr. Rife's reputation and his long record of work."*

Judge Kelly: *"Was the great secret the frequencies that would heal a cancer?"*

Rife's Attorney: *"Yes."*

Witness: *"Each disease has its own particular frequency."*

Judge Kelly: *"Could anyone do it?"*

AMA Attorney: *"No, a person who has not the key to these dials and bands could not."*

Rife's Attorney: *"Dr. Rife is not going to be a party to this fraud."*

Judge Kelly: *"We have here an instrument designed by Dr. Rife ... to benefit humanity, thoughtless of his own interests."*

—From the Court transcript of 1939 in which the American Medical Association sought to hijack Royal Rife's scientific discoveries.

The AMA boss who destroyed the unfolding new science of electro-medicine actually admitted in writing that he was motivated to kill electro-medicine because its existence threatened the chemotherapy and drug doctors' and industry interests. The cancer patients didn't concern him—he was totally oblivious to them and their suffering. He felt no obligation toward them. He simply wanted to own the new science. He apparently decided that if he couldn't own Royal Rife's discovery, then such a huge new approach to healing might obliterate him and his high position and the industry which backed him.

> *How is the individual physician who knows little or nothing of the physical basis of electricity and, in fact, who knows little or nothing of any physics at all ... to have any actual knowledge of these modalities?*

> *The introduction of unknown forces into the treatment of disease meant that physicians untrained in the basic sciences ... must begin anew their ... education, or yield their ... patients and their livelihood to those better informed.* Morris Fishbein, Fads and Quackery in Medicine, pages 299, 300.

"Or yield their patients"! The crucial motivation that drove the AMA to destroy the demonstrated cure for cancer! The total horror of that simple truth is staggering. Monstrous suffering and death resulted for millions and millions of people in the many decades after the court action of 1939 was long past.

In June, 1931 the national magazine *Popular Science* published an article that described the awesome, beautiful science of Royal Rife. A small portion follows. It should show 21st century readers how Royal Rife's achievements went far beyond what even 21st century scientists can now do.

> *Rife has devised a magnetic compass so delicate that it can be used to study the electricity and magnetism in living germs. He suggests that if the electrical make-up of dangerous germs is learned, it may someday be possible to destroy them in the human body by applying small doses of electricity.*

> *He showed me a quartz slide bearing several hundred typhus germs. "We have weighed them on extremely delicate balances ... The weight of these disease germs averages one-184-trillionth part of an ounce."*

The beast wasn't only the American Medical Association's criminal acts of 1938-1939. It also included health officials of the U.S. government health agencies FDA and NCI. For decades they had conspired to keep Royal Rife's demonstrated cure for cancer and entirely new field of science from being objectively researched and tested. Thus the beast was very much alive in modem form.

The following pertinent remarks provide insight into a terrible tragedy caused by conformism, ignorance, and a general bureaucratic reluctance to blow the whistle on a monstrous wrong.

> *There have already been many cancer cures, and all have been ruthlessly and systematically suppressed ... by the cancer establishment. The cancer establishment is the not-too-shadowy association of the American Cancer Society, the leading cancer hospitals, the National Cancer Institute, and the FDA ... The health field is the only field where inequity, conflict of interest, and gross public deception can remain unchecked.* Robert C. Atkins, M.D., Dr. Atkins Health Revolution.

> *The atrocities now being committed in the name of orthodox medicine, the suppression of life-giving scientific data, the needless loss of lives ... excessive suffering ... ultimately, these criminals ... will be brought to trial.* Dr. Bruce Halstead.

> *Bruce Halstead, M.D. ... was given a five-year jail sentence ... The crime? Dr. Halstead prescribed and sold to several of his cancer patients some herbal remedies.* Robert C. Atkins, M.D., Dr. Atkins Health Revolution.

> *The British Medical Association has acknowledged that alternative medicine is not quackery ... Encouraging results have been achieved ... in treatment for cancer ... A profession whose greatest experts could deceive themselves for so long should encompass novelty with humility.* Editorial, the London Times newspaper.

Beauty eventually will conquer the beast. Royal Rife's wonderful science will gain recognition and be revived. Future generations of scientists and doctors will cancel the madness, stupidity, and corruption of their predecessors. The new scientists and new doctors will unscramble the beautiful guidelines of Royal Rife and deliver anew his consistent and reliable cures for many diseases.

A last look at a few features of beauty complete this chapter.

> *If you have a biological treatment that really creates a good response in human cancer patients you know this without a randomized control trial ... You don't need it for a breakthrough.* Patrick McGrady, Jr., Cancer Scandal, a one-hour videotape.

> *The (particle) form ... is one phase in the life history of many, if not all bacteria. The bacterial forms do not produce cancer, but the (particle) form does.* The renowned Canadian cancer expert O. C. Gruner, who worked with Royal Rife for many years, from his book Study of Blood in Cancer.

> *The light used with Dr. Rife's microscope is polarized, that is, it is passed through crystals that stop all rays except those vibrating in one particular plane. By means of a double reflecting prism built into the instrument, it is possible to turn this plane of vibration in any desired direction, controlling illumination of the minute objects in the field very exactly.* Science News, December 12, 1931—supplement to Science magazine.

> *Rife ... believes that the chemical baths themselves destroy the very germs that science is trying to pin under the microscope. So he is evolving a new method that will do away with chemicals.*

> *...he expects within three minutes to place a perfectly normal, undoped slice of the diseased substance in position for examination.*

> *The possibilities of this process, once it is perfected, he believes, are boundless. Medical men ... may in this one step find an end to much of human suffering.* "Local Man Bares Wonders of Germ Life," San Diego Union newspaper, November 3, 1929.

... a frequency of light ... coordinates with the chemical constituents of the particles or micro-organisms under observation.

The variation of the light frequency is accomplished by use of a variable monochromatic beam of light that is tuned to coordinate with the chemical constituents of particle ... or micro-organism. Visibility of the particle ... is observed by use of the core beams from the patented Rife Microscope Lamps...

Rotation of the light beams in the quartz prisms controls the increase or decrease of the light frequency ... The control of the illumination ... is a most important factor in visualizing the ... pathogenic micro-organism. This cannot be accomplished by any conventional source of illumination. This points out why other research groups have failed to find the cancer (particle). The Smithsonian Institute Annual Report, U. S. government Printing Office, 1945, Publication 3776.

I put my hand on his stomach which was just one solid mass, just about what I could cover with my hand. It was absolutely solid! And I thought to myself, well, nothing can be done for that. However, they gave him a treatment with the ... frequencies and in the course of time over a period of six weeks to two months, to my astonishment, he completely recovered. James Couche, M. D., witness at the 100% successful 1934 cancer clinical trial using Royal Rife's revolutionary electro-medicine technology.

Chapter 8
Breast Cancer Basics: The Unscientific and the Unmentionable

The so-called medical authorities do not want certain disturbing truths about breast cancer revealed to the American public. Such an event would be very embarrassing for a select few in powerful positions of influence. Medical revolutions might happen as a result. The existing system of privileges along with the comfortable continuation of the status quo could be disrupted forever.

What if the very lucrative world of breast cancer treatment could be extinguished virtually overnight?

According to the cover story for *TIME* magazine dated October 15, 2007 by Kathleen Kingsbury:

> *In the U. S. an estimated $8.1 billion is spent to diagnose and treat breast cancer each year. (Page 37)*

> *Traditional chemotherapy may cost $20,000 or more. (Page 43)*

Breast cancer numbers are gruesome despite all the marches, various fund-raising events, and the gross amounts of money that are sent to researchers.

> *Breast cancer is the most common cancer in America's women. An estimated 178,480 new cases and 40,460 deaths from it are expected in the United States this year.* Marilynn Marchione, Associated Press newspaper service, September 6, 2007.

With the size of the breast cancer disaster now defined, the next step is to describe the unmentionable.

Royal Rife's wondrous, suppressed science has two surviving, recorded case histories of stunning breast cancer cures. Many others were probably lost when the son of a doctor who worked with Rife tossed all the records and case histories of his doctor father's dedicated, detailed work with Royal Rife's electro-medicine technology.

The 2007 research budget for just breast cancer according to the National Cancer Institute: $690 million. However, not a dime of it will be directed toward examining honestly and objectively the scientific principles that brought the incredible results which are about to be described.

In two separate interviews, one in the late 1950's, the second in the mid-1970's, the remarkable case of a breast cancer cure was told. The witness recalled how Dr. James Couche—who was present at the 1934 cancer clinical trial—used his own Rife instrument in 1938-1939 to accomplish what was then a consistent and reliable, scientifically-based healing.

> *Commander Harrison's wife had a breast carcinoma. The lower half of the breast was eaten away and she suffered intense agony all the time. But she didn't want to have it operated on because her husband had told her she would die of the metastasis and spreading it out in the bloodstream.*

> *Well, he (Dr. Couche) healed her up perfectly and her breasts*
> *took on the normal shape. When she realized that she had*
> *two normal breasts again, it was marvelous.* Ben Cullen,
> transcript of taped interview, October 15, 1959.

The quotation obviously leads to another glimpse of the racketeering-like, inexcusable policies, behavior, and cultures of the National Cancer Institute (NCI) and the Food and Drug Administration (FDA). These are the two federal government health agencies that have existed for decades with their figurative heads in the sand.

The story continues with a jump from the late 1930's Southern California location to 1969 (30 years later!) when a different physician—desperate because of a rapidly failing breast cancer patient of his—experimented with the taboo Royal Rife science and revolutionary therapy:

> *In 1969, a Salt Lake City surgeon operated on a patient,*
> *removing the left breast due to carcinoma. Six months later*
> *it had spread throughout her lymph system. He predicted the*
> *extent of her life expectancy was six more months at the*
> *most and possibly only three months. Through a friend he*
> *became aware of this experimental therapeutic modality and*
> *requested use of it, as he had nothing to lose in testing this*
> *system at this point. She was considered a terminal case. The*
> *end result has been a complete recovery. The case history is*
> *available.* From the "Marsh Papers."

Two separate, identical results. With geography and time providing the proof of independent replication—a key requirement of scientific validation principles.

What remains to be explained? The self-serving, corrupt system for breast cancer treatment must at least acknowledge these results and provide the technology with the opportunity to prove itself once again. To do anything less is socially irresponsible and reprehensible.

Why aren't women with breast cancer told about these easily verified and stunningly significant pieces of American medical history?

Because it might damage the credibility of approved treatments for breast cancer? At $20,000 per patient! Because huge amounts of money devoted to chemotherapy could vanish overnight? Because all sorts of "promising" research projects with millions of dollars in grant money might dry up?

> *We sincerely believe ... that BX is the primary cause of cancer ... We have used Koch's postulates in our methods of recovery ... this has been repeated hundreds of times proving to our own satisfaction that BX ... is the cause of malignancy.* Royal Rife, History of the Development of a Successful Treatment for Cancer. 1953.

And let us not forget those devastating court transcripts, more of which appear below.

> Judge Kelly: *"What is your profession?"*

> Royal Rife: *"I am a scientist engaged in research work, pathological and bacteriology."*

> Judge Kelly: *"When you perfected this instrument, you thought to manufacture and distribute these machines?"*

> Royal Rife: *"I wanted to get them out to the public, as many as I could."*
>
> —Court transcript, San Diego, 1939.

> *... beams of light are bent ... the minute granule, which is too small to absorb ordinary stain, is stained in its true chemical color.* San Diego Union newspaper, April 16, 1934.

Chapter 9
The Great Reluctance

The great reluctance by medical authorities to admit a massive wrongdoing by conventional, orthodox medicine in the Royal Rife matter is certainly understandable. But that does not make it right.

And the ongoing censorship or unspoken policy that keeps the resulting tragedy far away from public scrutiny certainly appears to be a classic cover-up response. This kind of predictable behavior has been seen countless times by the American public when some politician or public official is caught in a lie, scandal, misdeed, or criminal activity.

Still, this cover-up and silence is in a class by itself. With 10,000 Americans dying of cancer every week and cancer doctors in the nation's cancer centers not daring even to attack experimentally the microbe or "particle" that was shown by proven scientific methods, procedures, and principles to be the cause of cancer, then something is monstrously malevolent.

Nothing the medical authorities can say or invent as an excuse can possibly justify such a dereliction. It is an obscene malpractice and intense deceit that they condone. Attempts at absolving themselves and their profession reduce to transparent excuses.

There is an entire new field of science—classification of disease-causing microbes by color—that is being willfully obstructed or suppressed by medical officials and the always available skeptics. Why? Because open and honest scrutiny might bring Royal Rife's incredible scientific achievements to public attention. If the truth were revealed, an inevitable clobbering by a furious public would be the result.

Any honest, objective evaluation of the documented evidence will corroborate Royal Rife's discoveries and accomplishments. The untruthful facade smothering Royal Rife's history-making scientific discoveries can only be called one thing: criminal negligence. Sarcastic and uneducated denouncements by doctors and institutional public relations officers are daily compounding the already nightmarish proportions of this tragedy. The death count is so high—off the scales! Polite disagreements and sterile, intellectual bantering are totally inappropriate.

The color classification of micro-organisms and the painless, non-invasive electro-medicine therapy of this genuine new field of science were brought to yesterday's world by Royal Rife's scientific genius and dedication. And it was definitely done according to long recognized scientific principles and procedures.

> *Dr. Lewellys Barker of Johns Hopkins Medical School, one of the nation's foremost authorities on internal medicine, is visitor here ... Dr. Barker had just come from a visit with Dr. Royal Rife ... "He seems to be a mechanical genius," Dr. Barker said of Dr. Rife.* San Diego Sun newspaper, January 7, 1932.

> *... this special illumination reveals the filter-passing organism in characteristic individual colors ... no two kinds or forms of organisms have been found to have the same colors.* San Diego Evening Tribune newspaper, May 11, 1938.

I worked for the benefit of humanity ... I know it (the BX microbe) is the causative agent of cancer ... the 16 cases (in the clinical trial of 1934) ... you can imagine what they gave us. They came in ambulances twice a week ... photographs before and after. This board—six of the most outstanding men of the United States (5 physicians, 1 pathologist) ... Our board pronounced them (the cancer patients) clinically cured. Statement by Royal Rife describing the USC medical school-sponsored cancer clinical trial.

Rife's discoveries were not wanted and his cure for cancer and unique microscope that enabled living micro-organisms to be studied were not allowed into mainstream medical practice. After the American Medical Association failed in its criminal endeavor to steal them, the organization (and associated medical officials) did not relent but began working a number of dirty deals behind the scenes. They finally succeeded in keeping Dr. Rife's brilliant work suppressed and hidden.

... the ... brothers came down here (to San Diego) from Los Angeles at the ... direct request of the head physician of the AMA ... they said, "Well, we would like to buy into this corporation." Well, I said, "This happens to be a closed corporation" ... So one of the brothers acted a little bit belligerent and he said, " ... you ought to do something about it ... It would be good for you. It might be healthy for you." Statement by Ben Cullen, Rife's best friend, from a university professor's transcript of mid-1970s audio taped interviews with people who knew Dr. Royal and witnessed the AMA's initial attempt at grabbing Rife's breakthrough science.

Rife should have become a recognized, honored giant of 20th century medical science instead of an invisible vestige of history. Millions of people who died in agony might have lived many more healthy years, while an entirely new medicine was initiated for all humanity.

Having spent every dime I earned in my research for the benefit of mankind, I have ended up as a pauper, but I achieved the impossible and would do it again. A court affidavit filed by Royal Rife in the early 1960's.

Chapter 10
"CAM"—Blatant Hypocrisy

CAM is an abbreviation for "complementary and alternative medicine." It represents establishment medicine's inadequate attempt to channel the emerging alternative medicine movement into something it can control or at least something perceived as only an adjunct to traditional medicine. In truth, CAM is mostly a blunt instrument fraud—a deliberate obfuscation.

U.S. News & World Report magazine did a cover story on CAM in the magazine's January 21, 2008 issue. The story by Avery Comarow described how various CAM investigators at medical schools and leading hospitals were seeking to integrate CAM principles into mainstream medicine. However, predictably, it was all the "soft" alternatives that were being "tested" and "integrated" with traditional drug medicine—yoga, healing touch, massage, acupuncture, prayer, and so forth.

The Big, Dangerous, Science-based discoveries such as Royal Rife's demonstrated cures were not even considered. Apparently no honesty, daring, or outspoken outrage characterized any of these pretentious "experts" from the new CAM universe. But grant money for future investigations certainly had their attention! As if they—who knew nothing of the decades-long Rife suppression—suddenly were qualified to judge everything labeled "alternative."

> *A large pot of federal and foundation research funds—now close to $250 million per year helped turn ... sniffy attitude into solicitous attention.* Avery Comarow, "Embracing Alternative Care," U. S. News & World Report magazine. January 21, 2008, page 36.

It was extraordinary that the various experts—newly enmeshed in CAM programs at the medical schools, hospitals, and a variety of research facilities were so self-satisfied with their supposed expertise. Yet, in truth, they were both ignorant and cowardly. They actually thought themselves open-minded and well-informed. But they apparently had no knowledge whatsoever of Royal Rife's science or demonstrated cure for cancer—or if they did, they chose to avoid that dangerous territory.

The label "energy medicine" to the CAM tribe apparently meant anything except electro-medicine destroying the micro-organism that caused cancer! In their inexplicable world, "energy medicine" meant things like "healing touch" or "massaging human auras"! How ridiculous can it get?

They couldn't bring CAM into something as rigorously scientific as Royal Rife's carefully documented discoveries. The CAM folks apparently have no awareness that such a world exists—even though endless information and documentation is available, only waiting to be discovered and explored.

Instead, the CAM folks live in their own quaint little world isolated and restricted to interaction with each other and the various permitted "complementary" categories—yoga, acupuncture, healing touch, other kinds of approved "energy medicine" modalities, meditation, and so forth. All of it is safely non-threatening and can be permitted as tokenism that standard medicine can live with.

Still, the CAM crowd pretentiously imagine themselves to be on the cutting edge of medical virtue, curiosity, and intelligence. Of course, 10,000 Americans will continue to die of cancer every week as they slowly proceed to test their enshrined CAM therapies while seeking grant money.

And all of it promoted by Establishment Medicine's sweet double-talk. It catches the ignorant and unwary so easily. The white coat of authority worn by the newly emerging CAM organization (doctors, researchers, appointed government officials, and supposed civil servants) is unquestioned by the public at large.

The author of the CAM article conveniently ignores and mischaracterizes the blunders of mainstream medicine. Take, for example, Avery's reference to bone marrow transplant.

Some background information is in order: This 1990's scandal is fairly easy to trace and document. The bone marrow transplant program of that decade was based on a bogus study where the numbers had been cooked. An ambitious young doctor caused the incredible pain and suffering. He is now one of the leading lights of the cancer establishment. Naturally, such a disgusting truth is avoided by the mainstream press.

Even after the *Wall Street Journal* newspaper of November 17, 1994 did a front page exposure of the utter worthlessness of the financially ruinous, painful procedure, none of the elected "political guardians" in Washington, D.C. did anything to help the abused patients.

For the next 4½ years, the doctors and hospitals conspired in a greed-driven madness until finally, in March-April 1999, hard, factual clinical evidence irrefutably demonstrated that bone marrow transplant (a cute name used to disguise the high dose chemotherapy regimen at the core of the treatment) was not capable of the promised results.

Yet the CAM story's naive reporter wrote about the continuing use of the bone marrow transplant procedure as if it was a respected method that one or another CAM modality might help! It was such an incredible ignorance, assumption, and failure to do basic homework by this reporter.

The result is obvious—the public is definitely not as informed as they believe. Instead, the public is being tragically cheated.

Royal Rife's grand science and beautiful medicine waits patiently for its time of recognition and public exposure. If only one or a few of the new CAM tribe dares to demand the real, scientific stuff be brought into their light! Maybe the CAM tribe's learning curve with the real alternative medicine may also produce a pinch of courage. Maybe?

The great legal thinker and Supreme Court Justice Benjamin Cardozo long ago provided the solution to these horrible, censored, and suppressed wrongs: "Sunlight is the best disinfectant."

> *... 16 cases were treated at the clinic for many types of malignancy. After 3 months ... these so-called hopeless cases were signed off as clinically cured by the staff of five medical doctors.* Royal Rife, History of the Development of a Successful Treatment for Cancer.

> *Under the universal microscope disease organisms such as those of tuberculosis, cancer, sarcoma ... typhoid ... and others may be observed to succumb when exposed to certain lethal frequencies peculiar to each individual organism, and directed upon them by rays covering a wide range of waves.* Annual Report of the Board of Regents of The Smithsonian Institution ... for the year ended June 30, 1944. United States Government Printing Office, Washington, 1945, Pub 3776.

The misdeeds of oncologists and the cancer industry, and their total failure to honor their own Hippocratic Oath, eventually will produce an absolutely guaranteed future shift. Rebellions against the current Medical Establishment—with its censorship and self-interest mentality—are inevitable. The rebellions will come from both within the doctor ranks as well as from the outsiders—the patients and the genuine alternative medicine advocates. Some even will be waving the Royal Rife flag and carrying the new doctor tool of proven frequency instruments or ray tubes. Except these future Rife tools will be devices that produce consistent and reliable results—as did Royal Rife's original devices (not the numerous, shoddy products often fraudulently marketed by today's exploiters under the name "Rife instrument").

However and whenever the back-scratching, good old boy cronyism of Establishment Medicine's Club ends, it cannot be soon enough. Tragically, civilization and its geniuses have always had to endure ignorance in the slow turning of the wheel of progress.

> ... the gap between what people need or want and what they get from their government has grown to monstrous proportions ... People who desperately need help don't get it ... democracy has marked virtues that the centralist or authoritarian decision system lacks. Alvin Toffler, from the Foreword and Preface to Anticipatory Democracy, edited by Clement Bezold.

> A growing number of people are agitating for the right of patients to choose their own forms of medical treatment ... Complex legal issues involving constitutional and antitrust law with regard to treatment choice abound. Alan L. Glaser, "Alternative Healing and the Law," Medical Trial Technique Quarterly, Fall, 1994.

> Conflict of interest undermines the doctors claim to authority ... courts may ... identify and protect patient choice. Marjorie Schultz, The Yale Law Journal, Volume 95.

Western doctors ... largely ignoring the contributions of complementary philosophies will probably be seen in several centuries' time as being akin to the Ptolemaic view that the earth was the center of the universe. Richard Horton, The Lancet, March 6, 1993.

I will give no deadly medicine to anyone ... Into whatever house I enter, I will go into them for the benefit of the sick, and will abstain from every voluntary act of mischief and corruption. The Hippocratic Oath

APPENDICES

Appendix A
Law Matters: Justice Absent

When the law is no longer functioning, a Constitution-based Republic begins to disintegrate. Whatever the reason—fear, corruption, abuse of power—the result is the same. Justice dies. Citizens evade and deny. Bribes, payoffs, awful and selfish behavior flourishes, and other indications of societal deterioration take over. Freedom of speech withers. Various kinds of intimidation produce a reluctance to get involved. Simple solutions to obvious problems disappear or are quickly discounted.

There are issues developing that need to be addressed within the bureaucratic government health agencies. These issues belong to the Department of Justice. They involve government malfeasance, fraud, and criminal activity. They have long been festering. Crimes have been and continue to be committed.

Someday these years and years of systematic crimes by people within U.S. government health agencies will become ongoing front page news. The reader must be prepared for that eventual, momentous shift. Many scientists and high officials from the health agencies—in both Republican and Democratic administrations—will be switched suddenly to criminal classifications because of their misdeeds and subsequent grand jury indictments. Others will escape the jail dangers, but will continue to be identified through press exposures as having participated in deceits against the public interest.

"We didn't know" will become a commonly heard excuse, mantra, slogan, or echo. Such a phony alibi will not work with a lot of the public who finally get to read about and consider the facts and the documented evidence for themselves.

Justice may be absent now in the current environment. But it has a way of renewing itself periodically.

The full dimension of these problems, outrages, and illegal deeds has barely been exposed in this book, but the reader has been alerted to a dawn and a new day that will arrive—an awakening. And then Justice again will reign and show its appropriate hard essence.

> *The public good and the advantage of society are the purposes of establishing a government ... power cannot be arbitrary ... it must be exercised according to the established laws so that the people may know its duty and be secure within the shelter of laws.*

—Diderot's Encyclopedie (18th century France)

Appendix B
They Betrayed the Public Trust:
Rife's Purposely Lost Science

Faking is something doctors regularly do when admitting their ignorance concerning a huge medical subject would diminish their authority. They therefore audaciously fake and lie concerning the disturbing subject known as the documented Rife cancer cure. For the doctors to acknowledge that they have never investigated the evidence of an actual cure for cancer is a terrible truth that they would prefer not to explain. Especially when 10,000 Americans die from cancer every week while the expensive, medically approved treatments continue to be mostly abject failures. The situation is a ticking time bomb. And the public grows more resentful as the years of "research" and expertise in cancer "remedies" simply go nowhere. Just more of the same talk, same approaches, and same similarly educated "club" of cancer specialists protecting their own interests.

The doctors and scientists who work for the U.S. government health agencies are deeply implicated in the ongoing tragedy and cover-up of Royal Rife's historic, documented cure for cancer. They have blatantly betrayed the public trust.

The bureaucrats at the health agencies have the wildest excuses imaginable for their failure. However, all the excuses and rationales essentially reduce to keeping their positions within a comfortable, credentialed, bureaucratic fiefdom or insider's club. Outsiders are ignored, dismissed or never have legitimate, history-making discoveries objectively evaluated. Never!

To the distorted thinking of the government-employed cancer "experts" a cure for cancer must come from within their club and must be based on an appropriately designed, appropriately financed, and statistically approved project. Woe to those outsiders who attempt to bring anything original to desperate cancer patients.

And thus millions of Americans die of cancer while the cancer "experts" shamelessly promote the latest Big Pharma, expensive chemo concoctions.

The doctors and scientists rarely take chances and never do courageous whistle-blowing or muckraking exposures of despicable policies and deadly wrongs. This is absolutely a monstrous betrayal of the public trust.

Still, history has a way of exposing and changing rotten systems. Often a lightning bolt flash out of nowhere brings a view of an entirely new world, resulting in a giant leap for civilization!

But as long as self-serving mediocrities are put in positions of power, disaster and corruption are almost always inevitable. The doctors and scientists who have kept silent despite the documented proof of Royal Rife's cure for cancer deserve the American people's total contempt.

There exists a kind of bureaucratic insanity where the U.S. government health agencies' civil servants make assumptions that are just absolutely untrue. They continue to raid people who build and sell experimental Rife instruments simply because the regulations require such misdeeds. Regardless of the human cost in lives.

No examination or even superficial knowledge of Rife's incredible, documented scientific achievements apparently has ever made one of these bureaucrats ask, "What in God's name are we doing?"

After all, Royal Rife merely discovered an entirely new science where a specific germ's color identified the cause of a specific disease. Plus Rife painlessly, non-invasively destroyed the germ responsible for the disease. Rife was backed by top scientists from Johns Hopkins medical school, the Mayo Clinic, Northwestern, and McGill University in Montreal. An article concerning Rife's breakthrough cancer treatment appeared in the Smithsonian Institute's annual report. Science magazine reported on Rife's mind-boggling, technological discoveries and genius-level inventions.

Yet today's obstinate bureaucrats as well as the politically appointed directors of the health agencies continue to attack those courageous individuals seeking to replicate Rife's stunningly effective electro-medicine instruments.

The U.S. government bureaucrats—failing to do any serious preliminary investigation—deserve to be thought of as definite betrayers of the public trust. Why? Simply because they refused to examine easily obtained documented scientific evidence of Rife's accomplishments as well as easily obtained documented medical discoveries by Rife of grand historic importance.

Clearly, many government health employees have good intentions and contribute to the welfare of the public. However, the government culture has created a large number of health agency bureaucrats who are nothing more than impotent paper pushers, 2nd rate scientists, and mental midgets posing as health cops serving the American people.

Appendix C
Unmistakably Newsworthy

The press can't handle it. The story is too big, too awful, and too incredible to be true. And if true, it is too terrible to report to the American people. A long existing news blackout is maintained. The press corps avoids the subject instinctively. Occasionally a maverick reporter asks the awful question to a medical official who automatically lies. The insidious corruption of honesty and integrity within the doctor world spreads.

Undoubtedly the medical official would be much more selective in his choice of words if they were given during legal testimony under oath. Perjury fears will induce truth telling. But since doctors dismissing Royal Rife's documented achievement when asked by a curious reporter brings no legal threat, the doctors simply lie.

Thus the press gets routinely discouraged from investigating deeply whenever the subject of a proven, documented cure for cancer is discussed.

The conclusion thus is predictable—the potential story and the allegation are not newsworthy. Except why would people continuously lie on a subject so critically important? The simple answer is: because to report the story would cause a massive scientific and medical scandal. The world would be changed overnight. Those in major medical power positions would suffer a disruption that would end their cozy time at the medical heights.

Therefore the doctors routinely and shamelessly lie to the press. And the reporter then has the excuse that he wants in order to avoid investigating the evidence. The reporter races for the convenient escape hatch. The dangers associated with a huge newsworthy story thankfully disappear.

"Newsworthy" is a meaningless word when news blackout is policy. But the internet world surely worries the medical authorities and high U.S. government health officials because censorship is near impossible. The long taboo Rife cure for cancer—if exposed to the American people—could bring instant misery and a long overdue scorching public spotlight. The internet brings this occurrence closer each day. And it is unstoppable.

Appendix D
Poisonous Conduct

In 2006, the *New York Times* broke the story that many cancer doctors (oncologists) were taking payoffs from drug companies to push chemotherapy that didn't work on unsuspecting cancer patients. While America's famed "newspaper of record" *(New York Times)* courageously exposed a monstrous evil, the TV broadcasting world continued to avoid telling the larger American TV audience about the betrayal of the medical doctors' oaths.

But as the consistent, great journalist Ralph W. Moss reported in a *Townsend Letter for Doctors and Patients* magazine article (August/September, 2007) regarding the *New York Times'* scandal exposing revelation:

> *... Cancer patients will have no way to know whether a particular drug is being prescribed because it is the right and proper treatment, or because it lines the pockets of the doctor who gives it.*

Ralph W. Moss added to his bull's-eye declaration that other oncologists, by remaining silent were collectively contributing to a frightening wrong— "guilty of enabling their more greedy colleagues."

In blunt language, the cancer world of white-coated "professional" oncologists and so-called authorities had been revealed as derelict in their duty to put first the interests of the cancer patients, not the bank accounts of the cancer doctors.

A shameful truth of long duration (decades) regarding the basic failure of chemotherapy to cure cancer was slowly making its way to a breakout explosion into mainstream Americans' awareness. Unfortunately, the complete, ugly, fully exposed mess must rest with future generations to discover and reveal. However, documented evidence of the evil by both doctor associations and journalism groups is sufficient to convince reasonable people that both cancer doctors and journalists betrayed the public interest again and again.

This long censored, blasphemous behavior will remain in the history books of American medicine for future generations of doctors and the public.

It deserves an enduring condemnation. Such actions were monumentally immoral.

Some ugly facts that the American people are not told include these terrible yet easily-shown revelations by a highly qualified cancer specialist at a major cancer institute:

> *Since its formal launch more than thirty-five years ago, the war on cancer has been fighting many of the wrong battles with the wrong weapons and the wrong leaders.*
>
> *... war and poison gas provided the foundation for chemotherapy.*
>
> —Devra Davis, *The Secret History of the War on Cancer*, (2007), pp. xii, 201

A similar disgusting revelation by political commentator Dick Morris warning a slowly awakening public (echoes of Paul Revere!):

> *... Frequently, the pharmaceutical companies bribe doctors.*
>
> *... millions more to influence politicians so they won't curb their (drug companies) practices.*
>
> —Dick Morris and Eileen McGarn, *Outrage* (2007), pages 200, 203

According to a mid-May, 2008 Reuters news service article by Bill Berkret, the 2008 cost of cancer drugs is estimated to be $48 billion.

The drug companies love the profits derived from cancer treatment focused on chemotherapy.

The cancer experts employed by the U. S. government health agencies do not dare take even a small initiative in an honestly conducted, Rife-related, scientific investigation. Their own personal careers could be wrecked by such an initiative. So they just duck and evade while 10,000 Americans die each and every week because of this reluctance to risk by civil servants perpetrating a public service abomination.

Complementing the U. S. government health agencies' deceit and malfeasance are the sham actions of young attorneys at the U. S. Justice Department who seek personal legal advancement through prosecution of Rife-related activity by individuals or companies. These promotion-seeking attorneys have little understanding of or respect for justice's obligations. They thus go forth, seeking indictments from grand juries by failing to tell the citizen jury the entire truth regarding Dr. Rife's momentous, historically documented scientific and medical accomplishment.

A single grand juror's probing questions might derail an ambitious Assistant Attorney General's career goal! Therefore, relevant facts often are hidden from the grand jury when an indictment is sought against Rife activists in modern, 21st century times. Evidence is concealed by a Department of Justice attorney despite his or her oath.

The foundations of justice and the entire American legal system are obviously compromised and disgraced by such practices. Still, time eventually will deliver repercussions. And they won't be pretty for those who committed such offenses against the noble principles still secure at America's bedrock.

Those guilty of such abuse of legal power have been here alerted and advised. A reckoning will come. And there won't be many places to hide from the furies who guard the higher precepts of law.

Appendix E
Barry Lynes' Original Rife History Book
The Cancer Cure That Worked

PUBLISHER'S NOTE:

In 1987, Barry Lynes wrote the classic book on Rife history called *The Cancer Cure That Worked*. The book you now hold in your hands, *Rife's World of Electromedicine,* is the sequel, published in 2009. What is the difference between the two books? *The Cancer Cure That Worked* was primarily a biographical account of Royal Raymond Rife's life and work, including detailed, dated records of the events which occurred, and in-depth accounts from the people involved.

Rife's World of Electromedicine, on the other hand, is a bird's eye view, short summary of the same time period and events. Instead of a detailed biography, *Rife's World* is an expository piece that includes brief, targeted chapters addressing each aspect of Rife's era, utilizing piercing, specific, and direct quotes and excerpts from historical documents including magazine and newspaper articles, court transcripts, Rife's own statements, and the eye-witness accounts of those who were present during Rife's lifelong achievements.

In its more than 20 year lifespan, *The Cancer Cure That Worked* has sold over 50,000 copies and has been purchased by readers in more than 50 countries. As a testament to the value of the book, it should be noted that sales numbers for *The Cancer Cure That Work* have not slowed down at all since its original printing in 1987. The world has come to recognize *The Cancer Cure That worked* as the authoritative and historically accurate account of Royal Raymond Rife's life and scientific achievements.

If you enjoyed reading *Rife's World of Electromedicine*, then you should consider also reading *The Cancer Cure That Worked*, available from:

www.cancer-cure-that-worked.com

Or by phone at: (530) 541-7200

INDEX

BioMed Publishing Group
Product Catalog

Books & DVDs on
Rife Therapy &
Related Topics

To Order Products:
Call (530) 573-0190
Click www.BioMedPublishers.com

When Antibiotics Fail...

LYME DISEASE AND RIFE MACHINES

With Critical Evaluation of Leading Alternative Therapies

A book about how experimental frequency devices known as 'rife machines' have been used for over 15 years in private homes to successfully fight chronic Lyme Disease.

By Bryan Rosner
Edited by Michael Huckleberry

Foreword by Richard Loyd, Ph.D.
Coordinator of the Annual Rife International Health Conference

Contributors:
Marc Fett
Doug MacLean

Including:
Extensive
Internet
Resources

"Finally a valid, non-pharmaceutical treatment for this devastating illness." -Robert Andrew, M.D.

Book • $35

When Antibiotics Fail: Lyme Disease And Rife Machines, With Critical Evaluation Of Leading Alternative Therapies

By Bryan Rosner
Foreword by Richard Loyd, Ph.D.

There are enough books and websites about what Lyme disease is and which ticks carry it. But there is very little useful information for people who actually have a case of Lyme disease that is not responding to conventional antibiotic treatment. Lyme disease sufferers need to know their options, not how to identify a tick.

This book describes how experimental electromagnetic frequency devices known as rife machines have been used for over 15 years in private homes to fight Lyme disease. Also included are evaluations of more than 25 conventional and alternative Lyme disease therapies, including:

- Homeopathy
- IV and oral antibiotics
- Mercury detox.
- Hyperthermia / saunas
- Ozone and oxygen
- Samento®
- Colloidal Silver
- Bacterial die-off detox.

- Colostrum
- Magnesium supplementation
- Hyperbaric oxygen chamber (HBOC)
- ICHT Italian treatment
- Non-pharmaceutical antibiotics
- Exercise, diet and candida protocols
- Cyst-targeting antibiotics
- The Marshall Protocol®

Many Lyme disease sufferers have heard of rife machines, some have used them. But until now, there has not been a concise and organized source to explain how and why they have been used by Lyme patients. In fact, this is the first book ever published on this important topic.

The Foreword for the book is by Richard Loyd, Ph.D., coordinator of the annual Rife International Health Conference. The book takes a practical, down-to-earth approach which allows you to learn about*:

> "This book provides life-saving insights for Lyme disease patients."
>
> **- Richard Loyd, Ph.D.**

- Antibiotic treatment problems and shortcomings—why some people choose to use rife machines after other therapies fail.
- Hypothetical treatment schedules and sessions, based on the author's experience.
- The experimental machines with the longest track record: High Power Magnetic Pulser, EMEM Machine, Coil Machine, and AC Contact Machine.
- Explanation of the "herx reaction" and why it may indicate progress.
- The intriguing story that led to the use of rife machines to fight Lyme disease 20 years ago.
- Antibiotic categories and classifications, with pros and cons of each type of drug.
- Visit our website to read FREE EXCERPTS from the book!

Disclaimer: *Your treatment decisions must be made under the care of a licensed physician. Rife machines are not FDA approved and the FDA has not reviewed or approved of these books. The author is a layperson, not a doctor, and much of the content of these books is a statement of opinion based on the author's personal experience and research.*

Paperback book, 8.5 x 11", 203 pages, $35

13 Lyme Doctors Share Treatment Strategies!

In this new book, not one, but thirteen Lyme-literate healthcare practitioners describe the tools they use in their practices to heal patients from chronic Lyme disease. Never before available in book format!

Get Inside The Minds Of Top Lyme Doctors!

Insights Into Lyme Disease Treatment:
13 Lyme Literate Health Care Practitioners
Share Their Healing Strategies

By Connie Strasheim
Foreword by Maureen Mcshane, M.D.

If you traveled the country for appointments with 13 Lyme-literate health care practitioners, you would discover many cutting-edge therapies used to combat chronic Lyme disease. You would also spend thousands of dollars on hotels, plane tickets, and medical appointment fees—not to mention the time it would take to embark on such a journey.

Even if you had the time and money to travel, would the physicians have enough time to answer all of your questions? Would you even know which questions to ask?

In this long-awaited book, health care journalist and Lyme patient Connie Strasheim has done all the work for you. She conducted

Paperback • 443 Pages • $39.95

intensive interviews with 13 of the world's most competent Lyme disease healers, asking them thoughtful, important questions, and then spent months compiling their information into 13 organized, user-friendly chapters that contain the core principles upon which they base their medical treatment of chronic Lyme disease. The practitioners' backgrounds span a variety of disciplines, including allopathic, naturopathic, complementary, chiropractic, homeopathic, and energy medicine. All aspects of treatment are covered, from anti-microbial remedies and immune system support, to hormonal restoration, detoxification, and dietary/lifestyle choices.

PHYSICIANS INTERVIEWED:

- Steven Bock, M.D.
- Ginger Savely, DNP
- Ronald Whitmont, M.D.
- Nicola McFadzean, N.D.
- Jeffrey Morrison, M.D.
- Steven J. Harris, M.D.
- Peter J. Muran, M.D., M.B.A.
- Ingo D. E. Woitzel, M.D.
- Susan L. Marra, M.S., N.D.
- W. Lee Cowden, M.D., M.D. (H)
- Deborah Metzger, Ph.D., M.D.
- Marlene Kunold, "Heilpraktiker"
- Elizabeth Hesse-Sheehan, DC, CCN
- Visit our website to read a <u>FREE CHAPTER</u>!

Paperback book, 7 x 10", 443 pages, $39.95

DVD • $24.50

Rife International Health Conference Feature-Length DVD (93 Minutes)

Bryan Rosner's Presentation and Interview with Doug MacLean

The Official Rife Technology Seminar Seattle, WA, USA

If you have been unable to attend the Rife International Health Conference, this DVD is your opportunity to watch two very important Lyme-related presentations from the event:

Presentation #1: Bryan Rosner's Sunday morning talk entitled *Lyme Disease: New Paradigms in Diagnosis and Treatment - the Myths, the Reality, and the Road Back to Health.* (51 minutes)

Presentation #2: Bryan Rosner's interview with Doug MacLean, in which Doug talked about his experiences with Lyme disease, including the incredible journey he undertook to invent the first modern rife machine used to fight Lyme disease. Although Doug's journey as a Lyme disease pioneer took place 20 years ago, this was the first time Doug has ever accepted an invitation to appear in public. This is the only video available where you can see Doug talk about what it was like to be the first person ever to use rife technology as a treatment for Lyme disease. Now you can see how it all began. Own this DVD and own a piece of history! (42 minutes)

Lymebook.com has secured a special licensing agreement with JS Enterprises, the Canadian producer of the Rife Conference videos, to bring this product to you at the special low price of $24.50. Total DVD viewing time: 1 hour, 33 minutes. We have DVDs in stock, shipped to you within 3 business days.

Price Comparison (should you get the DVD?)

Cost of attending the recent Rife Conference (2 people):
Hotel Room, 3 Nights = $400
Registration = $340
Food = $150
Airfare = $600
Total = $1,490

Cost of the DVD, which you can view as many times as you want, and show to family and friends:
DVD = $24.50

**Bryan Rosner
Presenting on
Sunday Morning
In Seattle**

**DVD
93 Minutes
$24.50**

Amalgam Illness
diagnosis and treatment
What you can do to get better
How your doctor can help

Andrew Hall Cutler PhD PE

Do you or someone you know suffer from:

Allergies	Manic depressive disorder
Asthma	Rheumatoid arthritis
Autoimmune diseases	Juvenile arthritis
Amylotrophic lateral sclerosis	Lupus erythematosus
Ankylosing spondylitis	Multiple chemical sensitivities
Myasthenia gravis	Environmental illness
Parkinson's disease	Chronic fatigue
Alzheimer's disease	Fibromyalgia
Schizophrenia recurrent disorders	Sciatica
Borderline personality disorder	Gastritis
Anxiety	Irritable bowel syndrome
Panic attacks	Colitis
Attention deficit hyperactivity disorder	Crohn's disease
Learning disabilities	Sleep disorders
Endocrine problems	Anorexia nervosa
Depression	Bulimia
Obsessive-compulsive disorder	Tonic syndrome

Or other life beating problems your doctor can't do much about?

Have you heard about the controversy surrounding mercury from dental fillings and other sources? Did you know you might actually be mercury poisoned? Would you like to find out? Let a highly qualified research scientist explain:

- why we know millions of people have undiagnosed mercury poisoning
- why conventional medicine is slow to accept this
- how you can find out if your problems are usually due to mercury
- how to cure mercury poisoning and get rid of your health problems

Book • $35

Amalgam Illness, Diagnosis and Treatment: What You Can Do to Get Better, How Your Doctor Can Help

By Andrew Cutler, PhD

This book was written by a chemical engineer who himself got mercury poisoning from his amalgam dental fillings. He found that there was no suitable educational material for either the patient or the physician. Knowing how much people can suffer from this condition, he wrote this book to help them get well. With a PhD in chemistry from Princeton University and extensive study in biochemistry and medicine, Andrew Cutler uses layman's terms to explain how people become mercury poisoned and what to do about it. The author's research shows that mercury poisoning can easily be cured at home with over-the-counter oral chelators – this book explains how.

In the book you will find practical guidance on how to tell if you really have chronic mercury poisoning or some other problem. Proper diagnostic procedures are provided so that sick people can decide what is wrong rather than trying random treatments. If mercury poisoning is your problem, the book tells you how to get the mercury out of your body, and how to feel good while you do that. The treatment section gives step-by-step directions to figure out exactly what mercury is doing to you and how to fix it.

"Dr. Cutler uses his background in chemistry to explain the safest approach to treat mercury poisoning. I am a physician and am personally using his protocol on myself."

- Melissa Myers, M.D.

Sections also explain how the scientific literature shows many people must be getting poisoned by their amalgam fillings, why such a regulatory blunder occurred, and how the debate between "mainstream" and "alternative" medicine makes it more difficult for you to get the medical help you need.

This down-to-earth book lets patients take care of themselves. It also lets doctors who are not familiar with chronic mercury intoxication treat it. The book is a practical guide to getting well. Sections from the book include:

- Why worry about mercury poisoning?
- What mercury does to you – symptoms, laboratory test irregularities, diagnostic checklist.
- How to treat mercury poisoning easily with oral chelators.
- Dealing with other metals including copper, arsenic, lead, cadmium.
- Dietary and supplement guidelines.
- Balancing hormones during the recovery process.
- How to feel good while you are chelating the metals out.
- How heavy metals cause infections to thrive in the body.
- Politics and mercury.

This is the world's most authoritative, accurate book on mercury poisoning.

Paperback book, 8.5 x 11", 226 pages, $35

Hair Test Interpretation: Finding Hidden Toxicities

By Andrew Cutler, PhD

Hair tests are worth doing because a surprising number of people diagnosed with incurable chronic health conditions actually turn out to have a heavy metal problem; quite often, mercury poisoning. Heavy metal problems can be corrected. Hair testing allows the underlying problem to be identified – and the chronic health condition often disappears with proper detoxification.

Hair Test Interpretation: Finding Hidden Toxicities is a practical book that explains how to interpret **Doctor's Data, Inc.** and **Great Plains Laboratory** hair tests. A step-by-step discussion is provided, with figures to illustrate the process and make it easy. The book gives examples using actual hair test results from real people.

Hair Test Interpretation
Finding Hidden Toxicities
Andrew Hall Cutler, Ph. D., P. E.

Toxicity Causes Health Problems

Book • $35

One of the problems with hair testing is that both conventional and alternative health care providers do not know how to interpret these tests. Interpretation is not as simple as looking at the results and assuming that any mineral out of the reference range is a problem mineral.

Interpretation is complicated because heavy metal toxicity, especially mercury poisoning, interferes with mineral transport throughout the body. Ironically, if someone is mercury poisoned, hair test mercury is often low and other minerals may be elevated or take on unusual values. For example, mercury often causes retention of

"This new book of Andrew's is the definitive guide in the confusing world of heavy metal poisoning diagnosis and treatment. I'm a practicing physician, 20 years now, specializing in detoxification programs for treatment of resistant conditions. It was fairly difficult to diagnose the heavy metal conditions before I met Andrew Cutler and developed a close relationship with him while reading his books. In this book I found his usual painful attention to detail gave a solid framework for understanding the complexity of mercury toxicity as well as the less common exposures. You really couldn't ask for a better reference book on a subject most researchers and physicians are fumbling in the dark about."
- Dr. Rick Marschall

arsenic, antimony, tin, titanium, zirconium, and aluminum. An inexperienced health care provider may wrongfully assume that one of these other minerals is the culprit, when in reality mercury is the true toxicity.

So, as you can see, getting a hair test is only the first step. The second step is figuring out what the hair test means. Andrew Cutler, PhD, is a registered professional chemical engineer with years of experience in biochemical and healthcare research. This clear and concise book makes hair test interpretation easy so that you know which toxicities are causing your health problems.

Paperback book, 8.5 x 11", 298 pages, $35

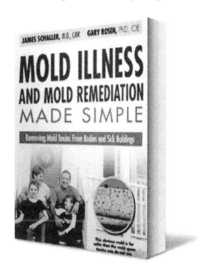

Book • $32.95

Mold Illness and Mold Remediation Made Simple: Removing Mold Toxins from Bodies and Sick Buildings

By James Schaller, M.D. and Gary Rosen, Ph.D.

Indoor mold toxins are much more dangerous and prevalent than most people realize. Visible mold in and around your house is far less dangerous than the mold you cannot see. Indoor mold toxicity, in addition to causing its own unique set of health problems and symptoms, also greatly contributes to the severity of most chronic illnesses.

In this book, a top physician and experienced contractor team up to help you quickly recover from indoor mold exposure. This book is easy to read with many color photographs and illustrations.

Dr. Schaller is a practicing physician in Florida who has written more than 15 books. He is one of the few physicians in the United States successfully treating mold toxin illness in children and adults.

Dr. Rosen is a biochemist with training under a Nobel Prize winning researcher at UCLA. He has written several books and is an expert in the mold remediation of homes. Dr. Rosen and his family are sensitive to mold toxins so he writes not only from professional experience, but also from personal experience.

Together, the two authors have certification in mold testing, mold remediation, and indoor environmental health. This book is one of the most complete on the subject, and includes discussion of the following topics:

- Potential mold problems encountered in new homes, schools, and jobs.
- Diagnosing mold illness.
- Mold as it relates to dryness and humidity.
- Mold toxins and cancer treatment.
- Mold toxins and relationships.
- Crawlspaces, basements, attics, home cleaning techniques, and vacuums.
- Training your eyes to discern indoor mold.
- Leptin and obesity.
- Appropriate/inappropriate air filters and cleaners.
- How to handle old, musty products, materials and books, and how to safely sterilize them.
- A description of various types of molds, images of them, and their relative toxicity.
- Blood testing and how to use it to find hidden health problems.
- The book is written in a friendly, casual tone that allows easy comprehension and information retention.

> "A concise, practical guide on dealing with mold toxins and their effects."
>
> **- Bryan Rosner**

Many people are affected by mold toxins. Are you? If you can find a smarter or clearer book on this subject, buy it!

Paperback book, 8.5 x 11", 140 pages, $32.95

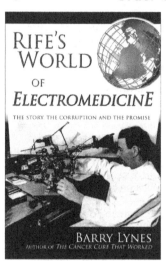

RIFE'S WORLD OF ELECTROMEDICINE
THE STORY THE CORRUPTION AND THE PROMISE

BARRY LYNES
AUTHOR OF *THE CANCER CURE THAT WORKED*

Book • $17.95

Rife's World of Electromedicine: The Story, the Corruption and the Promise
By Barry Lynes

The cause of cancer was discovered in the early 1930's. It was a virus-sized, mini-bacteria or "particle" that induced cells to become malignant and grow into tumors. The cancer microbe or particle was given the name BX by the brilliant scientist who discovered it: Royal Raymond Rife.

Laboratory verification of the cause of cancer was done hundreds of times with mice in order to be absolutely certain. Five of America's most prominent physicians helped oversee clinical trials managed by a major university's medical school.

Sixteen cancer patients were brought by ambulance twice a week to the clinical trial location in La Jolla, California. There they were treated with a revolutionary electromedicine that painlessly, non-invasively destroyed only the cancer-causing microbe or particle named BX. After just three months of this therapy, all patients were diagnosed as clinically cured. Later, the therapy was suppressed and remains so today.

In 1987, Barry Lynes wrote the classic book on Rife history (*The Cancer Cure That Worked*, see catalog page 14). *Rife's World* is the sequel, published in 2009.

Paperback book, 5.5 x 8.5", 90 pages, $17.95

Physicians' Desk Reference (PDR) Books (opposing page)

Most people have heard of *Physicians' Desk Reference* (PDR) books because, for over 60 years, physicians and researchers have turned to PDR for the latest word on prescription drugs.

THOMSON

You may not know that Thomson Healthcare, publisher of PDR, offers PDR reference books not only for drugs, but also for herbal and nutritional supplements. No available books come even close to the amount of information provided in these PDRs—*PDR for Herbal Medicines* weighs 5 lbs and has over 1300 pages, and *PDR for Nutritional Supplements* weighs over 3 lbs and has more than 800 pages.

> "I relied heavily on the PDRs during the research phase of writing my books. Without them, my projects would have greatly suffered."
> **- Bryan Rosner**

We carry all three PDRs. Although PDR books are typically used by physicians, we feel that these resources are also essential for people interested in or recovering from chronic disease. For the supplements, herbs, and drugs included in the books, you will find the following information: Pharmacology, description and method of action, available trade names and brands, indications and usage, research summaries, dosage options, history of use, pharmacokinetics, and much more! Worth the money for years of faithful use.

PDR for Nutritional Supplements   *2nd Edition!*

This PDR focuses on the following types of supplements:

- Vitamins
- Minerals
- Amino acids
- Hormones
- Lipids
- Glyconutrients
- Probiotics
- Proteins
- Many more!

"In a part of the health field not known for its devotion to rigorous science, [this book] brings to the practitioner and the curious patient a wealth of hard facts."

- Roger Guillemin, M.D., Ph.D., Nobel Laureate in Physiology and Medicine

Book • $69.50

The book also suggests supplements that can help reduce prescription drug side effects, has full-color photographs of various popular commercial formulations (and contact information for the associated suppliers), and so much more! Become educated instead of guessing which supplements to take.

Hardcover book, 11 x 9.3", 800 pages, $69.50

PDR for Herbal Medicines   *4th Edition!*

PDR for Herbal Medicines is very well organized and presents information on hundreds of common and uncommon herbs and herbal preparations. Indications and usage are examined with regard to homeopathy, Indian and Chinese medicine, and unproven (yet popular) applications.

In an area of healthcare so unstudied and vulnerable to hearsay and hype, this scientifically referenced book allows you to find out the real story behind the herbs lining the walls of your local health food store.

Use this reference before spending money on herbal products!

Book • $69.50

Hardcover book, 11 x 9.3", 1300 pages, $69.50

PDR for Prescription Drugs   *Current Year's Edition!*

With more than 3,000 pages, this is the most comprehensive and respected book in the world on over 4,000 drugs. Drugs are indexed by both brand and generic name (in the same convenient index) and also by manufacturer and product category. This PDR provides usage information and warnings, drug interactions, plus a detailed, full-color directory with descriptions and cross references for the drugs. A new format allows dramatically improved readability and easier access to the information you need now.

Book • $99.50

Hardcover book, 12.5 x 9.5", 3533 pages, $99.50

Book • $22.95

Sauna Therapy for Detoxification and Healing

By Lawrence Wilson, MD

This book provides a thorough yet articulate education on sauna therapy. It includes construction plans for a low-cost electric light sauna. The book is well referenced with an extensive bibliography.

Sauna therapy, especially with an electric light sauna, is one of the most powerful, safe and cost-effective methods of natural healing. It is especially important today due to extensive exposure to toxic metals and chemicals.

Fifteen chapters cover sauna benefits, physiological effects, protocols, cautions, healing reactions, and many other aspects of sauna therapy.

Dr. Wilson is an instructor of Biochemistry, Hair Mineral Analysis, Sauna Therapy and Jurisprudence at various colleges and universities including Yamuni Institute of the Healing Arts (Maurice, LA), University of Natural Medicine (Santa Fe, NM), Natural Healers Academy (Morristown, NJ), and Westbrook University (West Virginia). His books are used as textbooks at East-West School of Herbology and Ohio College of Natural Health. Go to www.LymeBook.com for free book excerpts!

Paperback book, 8.5 x 11", 167 pages, $22.95

Book • $19.95

Over 50,000 Copies Sold!

The Cancer Cure That Worked: Fifty Years of Suppression

At Last You Can Read... The Rife Report

By Barry Lynes

Investigative journalism at its best. Barry Lynes takes readers on an exciting journey into the life work of Royal Rife. **In 2009, we became the official publisher of this book. Call or visit us online for wholesale terms.**

"A fascinating account..." **-Alan Cantwell, MD**

"This book is superb." **-Florence B. Seibert, PhD**

"Barry Lynes is one of the greatest health reporters in our country. With the assistance of John Crane, longtime friend and associate of Roy Rife, Barry has produced a masterpiece..." **-Roy Kupsinel, M.D., editor of *Health Consciousness Journal***

Paperback book, 5 x 8", 169 pages, $19.95

ife Video Documentary
-Part DVD Set, Produced by
ero Zero Two Productions

Must-Have DVD set for your ife technology education!

In 1999, a stack of forgotten audio pes was discovered. On the tapes ere the voices of several people at the enter of the events which are the ubject of this documentary: a evolutionary treatment for cancer and a ractical cure for infectious disease.

The audio tapes were over 40 years ld. The voices on them had almost ded, nearly losing key details of erhaps the most important medical ory of the 20th Century.

But due to the efforts of the innaman Foundation, the faded tapes ave been restored and the voices on em recovered. So now, even though e participants have all passed away...

...they can finally tell their story.

2-part DVD Set • $39.95

"These videos are great. We show them at the Annual Rife International Health Conference."
-Richard Loyd, Ph.D.

"A mind-shifting experience for those of us indoctrinated with a conventional view of biology."
-Townsend Letter for Doctors and Patients

In the summer of 1934 at a special medical clinic in La Jolla, California, sixteen patients withering from terminal disease were given a new lease on life. It was the first controlled application of a new electronic treatment for cancer: the Beam Ray Machine.

Within ninety days all sixteen patients walked away from the clinic, signed-off by the attending doctors as cured.

What followed the incredible success of this revolutionary treatment was not a welcoming by the scientific community, but a sad tale of its ultimate suppression.

The Rise and Fall of a Scientific Genius documents the scientific ignorance, fficial corruption, and personal greed directed at the inventor of the Beam Ray achine, Royal Raymond Rife, forcing him and his inventions out of the spotlight and to obscurity. **Just converted from VHS to DVD and completely updated.**

ncludes bonus DVD with interviews and historical photographs! Produced in Canada.

Visit our website today to watch a FREE PREVIEW CLIP!

2 DVD-set, including bonus DVD, $39.95

Book • $25.95

The Lyme-Autism Connection: Unveiling the Shocking Link Between Lyme Disease and Childhood Developmental Disorders

By Bryan Rosner & Tami Duncan

Did you know that Lyme disease may contribute to the onset of autism?

This book is an investigative report written by Bryan Rosner and Tami Duncan. Duncan is the co-founder of the *Lyme Induced Autism (LIA) Foundation*, and her son has an autism diagnosis.

Tami Duncan, Co-Founder of the Lyme Induced Autism (LIA) Foundation

Awareness of the Lyme-autism connection is spreading rapidly, among both parents and practitioners. *Medical Hypothesis*, a scientific, peer-reviewed journal published by Elsevier, recently released an influential study entitled *The Association Between Tick-Borne Infections, Lyme Borreliosis and Autism Spectrum Disorders*. Here is an excerpt from the study:

> "Chronic infectious diseases, including tick-borne infections such as Borrelia burgdorferi, may have direct effects, promote other infections, and create a weakened, sensitized and immunologically vulnerable state during fetal development and infancy, leading to increased vulnerability for developing autism spectrum disorders. An association between Lyme disease and other tick-borne infections and autistic symptoms has been noted by numerous clinicians and parents."

> —Medical Hypothesis Journal.
> Article Authors: Robert C. Bransfield, M.D., Jeffrey S. Wulfman, M.D., William T. Harvey, M.D., Anju I. Usman, M.D.

Nationwide, 1 out of 150 children are diagnosed with Autism Spectrum Disorder (ASD), and the LIA Foundation has discovered that many of these children test positive for Lyme disease/Borrelia related complex—yet most children in this scenario never receive appropriate medical attention. This book answers many difficult questions: How can infants contract Lyme disease if autism begins before birth, precluding the opportunity for a tick bite? Is there a statistical correlation between the incidences of Lyme disease and autism worldwide? Do autistic children respond to Lyme disease treatment? What does the medical community say about this connection? Do the mothers of affected children exhibit symptoms? **Find out in this book.**

Paperback book, 6x9", 287 pages, $25.95

Renegade Patient: The No-Nonsense, Practical Guide to Getting the Health Care You Need   By Tedde Rinker, D.0.

Includes Patient Templates & Medical Forms

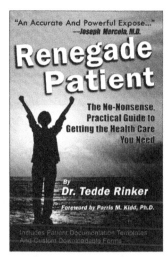

"An Accurate And Powerful Expose..."
—Joseph Mercola, M.D.

Renegade Patient

The No-Nonsense, Practical Guide to Getting the Health Care You Need

By Dr. Tedde Rinker

Foreword by Parris M. Kidd, Ph.D.

Includes Patient Documentation Templates And Custom Downloadable Forms

Stop! Before you pick up the phone to make an appointment with your local Lyme-literate doctor, take a few deep breaths, and ask yourself: "Am I an empowered patient who knows what I need from my physician and how to get it? Am I in control of my health care, and if not, who is?"

Why do you need to ask these questions? Unfortunately, many patients are not in control of their own health care. Physicians, politicians, pharmaceutical and insurance companies, and other third-party organizations are usually in the driver's seat—especially when it comes to controversial Lyme disease.

Book • $22.95

Dr. Rinker, an experienced osteopathic physician practicing medicine in Redwood City, California, believes that all patients should become empowered and responsible, equipped with the necessary education and knowledge to navigate the maze of modern medical services. This book includes tools for dealing with all aspects of the medical industry, from insurance companies and physicians' offices to requesting your medical chart and monitoring treatment progress. **This is a hands-on workbook with document templates and forms that you can actually use.**

Paperback book, 6x9", 245 pages, $22.95

Our catalog has space limitations, but our website does not! Visit www.LymeBook.com to see even more exciting products.

Don't Miss These New Books & DVDs, Available Online:
- Babesia Update 2009, by James Schaller, M.D.
- Marshall Protocol 5-DVD Set
- Cure Unknown, by Pamela Weintraub
- The Experts of Lyme Disease, by Sue Vogan
- The Lyme Disease Solution, by Ken Singleton, M.D.
- **Lots of Free Chapters and Excerpts Online!**

Don't use the internet? No problem, just call (530) 573-0190.

Book and Audio CD • $24.50

Outsmart Your Cancer: Alternative Non-Toxic Treatments That Work   By Tanya Harter Pierce

Why BLUDGEON cancer to death with common conventional treatments that can be toxic and harmful to your entire body?

When you OUTSMART your cancer, only the cancer cells die — NOT your healthy cells! *OUTSMART YOUR CANCER: Alternative Non-Toxic Treatments That Work* is an easy guide to successful non-toxic treatments for cancer that you can obtain right now! In it, you will read real-life stories of people who have completely recovered from their advanced or late-stage lung cancer, breast cancer, prostate cancer, kidney cancer, brain cancer, childhood leukemia, and other types of cancer using effective non-toxic approaches.

Plus, *OUTSMART YOUR CANCER* is one of the few books in print today that gives a complete description of the amazing formula called "Protocel," which has produced incredible cancer recoveries over the past 20 years. **A supporting audio CD is included with this book**. Pricing = $19.95 book + $5.00 CD.

Paperback book, 6 x 9", 437 pages, with audio CD, $24.95

Hardcover Book • $169.95

Cell Wall Deficient Forms: Stealth Pathogens

By Lida Mattman, Ph.D.

This is one of the most influential infectious disease textbook of the century. Dr. Mattman, who earned a Ph.D. in immunology from Yale University, describes her discovery that a certain type of pathogen lacking a cell wall is the root cause of many of today's "incurable" and mysterious chronic diseases. Dr. Mattman's research is the foundation of our current understanding of Lyme disease, and her work led to many of the Lyme protocols used today (such as the Marshall Protocol, as well as modern LLMD antibiotic treatment strategy). Color illustrations and meticulously referenced breakthrough principles cover the pages of this book. A must have for all serious students of chronic, elusive infectious disease.

Hardcover book, 7.5 x 10.5", 416 pages, $169.95

DVD • $24.50

Richard Loyd, Ph.D., presents at the Rife International Health Conference in Seattle

Watch this DVD to gain a better understanding of the technical details of rife technology.

Dr. Loyd, who earned a Ph.D. in nutrition, has researched and experimented with numerous electrotherapeutic devices, including the Rife/Bare unit, various EMEM machines, F-Scan, BioRay, magnetic pulsers, Doug Machine, and more. Dr. Loyd also has a wealth of knowledge in the use of herbs and supplements to support Rife electromagnetics.

By watching this DVD, you will discover the nuts and bolts of some very important, yet little known, principles of rife machine operation, including:

- Gating, sweeping, session time
- Square vs. sine wave
- DC vs. AC frequencies
- Duty cycle
- Octaves and scalar octaves

- Voltage variations and radio frequencies
- Explanation of the spark gap
- Contact vs. radiant mode
- Stainless vs. copper contacts
- A unique look at various frequency devices

DVD, 57 minutes, $24.50

The Lyme Disease Annual Report, by Bryan Rosner and Contributing Writers (2008)

Contributing writers include: **James Schaller, M.D., Richard Brand, M.D., Sue Vogan, Ginger Savely, FNP-C, Tami Duncan, Susan Williams, and Richard Loyd, Ph.D.** This informative report covers many topics, including glyconutrient supplementation, updates on rife machine treatment planning and machine options, evidence supporting the existence of chronic Lyme disease as a real medical condition, statistics

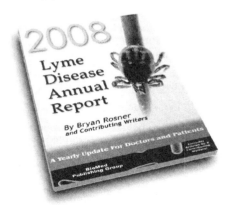

Paperback Book • $22.95

indicating the presence of Lyme disease on all continents of the planet, and much more. **Bonus! Free of charge: 9-page E-Book entitled "4 Immune-Supporting Supplements Every Lyme Sufferer Needs To Know About—And Where To Buy Them"** (instant download at checkout on our website).

Paperback, 168 pages, $~~22.95~~ *CLOSEOUT SALE: $14.95*
With Free E-Book, Downloadable Upon Checkout On Our Website!

The Rife Handbook of Frequency Therapy, With a Holistic Health Primer

New Revised Edition! By Nenah Sylver, PhD

This is the most complete, authoritative Rife technology handbook in the world. A hardcover book, it weighs over 2 lbs. and has more than 730 pages. A broad range of practical, hands-on topics are covered:

- New Revised Edition released in 2009 is twice as long as original book! Now with a complete index!
- Royal Raymond Rife's life, inventions, and relationships. Recently discovered data explaining how Rife's original machines worked.

Book • $99.95

- Frequently Asked Questions about Rife sessions and equipment, with extensive session information.
- Ground-breaking information on strengthening and supporting the body.
- A 200-page, cross-referenced Frequency Directory including hundreds of health conditions.
- Bibliography, Three Appendices, Historical Photos, Complete Index, AND MUCH MORE!
- DVD available at www.LymeBook.com with author's recent Rife Conference presentation.

Hardcover book, 8.5 x 11", 730 pages, $99.95

Ordering is Easy!

Phone: Toll Free (866) 476-7637
Online: www.LymeBook.com

Detailed product information and secure online ordering is available on our website. Bulk orders to bookstores, health food stores, or Lyme disease support groups – call us for wholesale terms.

Do you have a book inside you? Submit your book proposal online at: www.lymebook.com/submit-book-proposal.

Join Lyme Community Forums at: www.lymecommunity.com.

Get paid to help us place our books in your local health food store. Learn more: www.lymebook.com/local-store-offer.

DISCLAIMER

This disclaimer is in reference to all books, DVDs, websites, flyers, and catalogs published by Bryan Rosner, DBA BioMed Publishing Group.

Our materials are for informational and educational purposes only. They are not intended to prevent, diagnose, treat, or cure disease. Some of the treatments described are not FDA-Approved. Bryan Rosner is a layperson, not a medical professional, and he is not qualified to dispense medical advice.

These books and DVDs are not intended to substitute for professional medical care. Do not postpone receiving care from a licensed physician. Please read our full disclaimer online at: www.lymebook.com/homepage-disclaimer.pdf.

Made in the USA
Las Vegas, NV
17 March 2024